Relational Contracting for Construction Excellence

Spon Research

publishes a stream of advanced books for built environment researchers and professionals from one of the world's leading publishers. The ISSN for the Spon Research programme is ISSN 1940-7653 and the ISSN for the Spon Research E-book programme is ISSN 1940-8005

Free-Standing Tension Structures
From tensegrity systems to cable-strut systems
978-0-415-33595-9
W. B. Bing

Performance-Based Optimization of Structures
Theory and applications
978-0-415-33594-2
Q.Q. Liang

Microstructure of Smectite Clays & Engineering Performance
978-0-415-36863-6
R. Pusch & R. Yong

Procurement in the Construction Industry
The impact and cost of alternative market and supply processes
978-0-415-39560-1
W. Hughes et al.

Communication in Construction Teams
978-0-415-36619-9
S. Emmitt & C. Gorse

Concurrent Engineering in Construction Projects
978-0-415-39488-8
C. Anumba, J. Kamara & A.-F. Cutting-Decelle

People and Culture in Construction
978-0-415-34870-6
A. Dainty, S. Green & B. Bagilhole

Very Large Floating Structures
978-0-415-41953-6
C.M. Wang, E. Watanabe & T. Utsunomiya

Tropical Urban Heat Islands
Climate, Buildings and Greenery
978-0-415-41104-2
N.H. Wong & C. Yu

Innovation in Small Construction Firms
978-0-415-39390-4
P. Barrett, M. Sexton & A. Lee

Construction Supply Chain Economics
978-0-415-40971-1
K. London

Employee Resourcing in the
Construction Industry
978-0-415-37163-6
A. Raiden, A. Dainty & R. Neale

Managing Knowledge in the
Construction Industry
978-0-415-46344-7
A. Styhre

Collaborative Information
Management in Construction
978-0-415-48422-0
G. Shen, A. Baldwin and P. Brandon

Containment of High Level
Radioactive and Hazardous Solid
Wastes with Clay Barriers
978-0-415-45820-7
R.N. Yong, R. Pusch and M. Nakano

Performance Improvement in
Construction Management
978-0-415-54598-3
B. Atkin and J. Borgbrant

Relational Contracting for
Construction Excellence
Principles, Practices and Case studies
978-0-415-46669-1
A.P. Chan, D.W. Chan and J.F. Yeung

Forthcoming:

Organisational Culture in the
Construction Industry
978-0-415-42594-0
V. Coffey

Relational Contracting for Construction Excellence

Principles, Practices and Case Studies

Albert P. Chan, Daniel W. Chan and John F. Yeung

Routledge
Taylor & Francis Group

LONDON AND NEW YORK

First published 2010 by Spon Press

Published 2015 by Routledge
2 Park Square, Milton Park, Abingdon, Oxfordshire OX14 4RN
711 Third Avenue, New York, NY 10017

First issued in paperback 2015

Routledge is an imprint of the Taylor and Francis Group, an informa business

© 2010 Albert P. Chan, Daniel W. Chan and John F. Yeung

Typeset in Sabon by
HWA Text and Data Management, London

This publication presents material of a broad scope and applicability. Despite stringent efforts by all concerned in the publishing process, some typographical or editorial errors may occur, and readers are encouraged to bring these to our attention where they represent errors of substance. The publisher and author disclaim any liability, in whole or in part, arising from information contained in this publication. The reader is urged to consult with an appropriate licensed professional prior to taking any action or making any interpretation that is within the realm of a licensed professional practice.

British Library Cataloguing in Publication Data
A catalogue record for this book is available from the British Library

Library of Congress Cataloging-in-Publication Data
Chan, Albert P.
    Relational contracting for construction excellence : principles, practices and case studies / Albert P. Chan, Daniel W. Chan and John F. Yeung.
        p. cm. – (Spon research)
    Includes bibliographical references and index.
    1. Construction industry. 2. Customer relations. 3. Contracts. I. Chan, Daniel W. II. Yeung, John F. III. Title.
    HD9715.A2C47 2009
    624.068'4–dc22
                                                        2009002374

ISBN 13: 978-1-138-99719-6 (pbk)
ISBN 13: 978-0-415-46669-1 (hbk)

Printed and bound by CPI Group (UK) Ltd, Croydon, CR0 4YY

Contents

Figures

Tables

Foreword

Professor Ko Jan-ming
Vice President, The Hong Kong Polytechnic University

The construction industry has long been suffering from an adversarial relationship among the key parties, which is the result of ineffective communication and a lack of cooperation and mutual trust. Relational contracting, which covers a wide range of systems such as partnering, alliancing, public-private partnership, and joint venture, is a very topical subject attempting to overcome these hurdles.

There is a great need for a definitive work which sets out the many and diverse aspects of relational contracting. This book consolidates the research of the authors and explains in a systematic and coherent manner the various aspects involved in relational contracting. It starts with the fundamental concepts of relational contracting, followed by its historical development in major countries. Other important aspects such as benefits, difficulties, critical success factors and key performance indicators for implementing relational contracting are also addressed. The book highlights the explicit links between research findings and their practical applications and a number of real-life case studies from the UK, USA, Australia, and Hong Kong are provided.

I am therefore very pleased to recommend this book not only to the academic community but also to professionals who have dealings in relational contracting. This book provides an invaluable reference that neither you nor I should miss.

July 2009

Foreword

Thomas Ho
Chief Executive, Gammon Construction Limited

The Hong Kong construction industry is witnessing rapid changes in trading dynamics. With the onset of major infrastructure projects in the pipeline, closer regional collaboration within the Pearl River Delta and challenges ahead for sustainability and climate change, there will be an increasing demand for a collaborative and integral approach to tapping the expertise and networks among various parties in the supply chain, including developers, architects, contractors, consultants, suppliers and subcontractors. No single entity will have the adequate capital and power to provide the total solution required for some of the mega opportunities.

Even in today's business environment, construction projects are becoming more complex and sophisticated than ever. We find ourselves in the process of shifting from managing contracts to managing relationships to achieve win-win outcomes – from initial planning to building design and construction, and throughout the life cycle of the development. Relational Contracting which builds upon a cooperative, trust-based relationship and partnering spirit among industry players, is playing a key role in complex projects. Over the years, it has proven a sound model, offering significant benefits in terms of both efficiency and effectiveness.

It is delightful to see this book going live, showcasing various forms of Relational Contracting – partnering, alliancing, public-private partnership and joint venture. It also illustrates a series of real-life case studies in the US, UK, Australia and Hong Kong, providing useful facts, lessons learnt and benchmarks to researchers, practitioners and learners. The success stories in Hong Kong also demonstrate that the territory's construction industry is moving beyond the basics and toward state of the art.

In this book, you will find that team spirit and partnering are really the crux of project success. When individual problems become team problems, and all parties encourage and support each other to overcome hurdles, we are halfway through. It also highlights the importance of mutual trust and sharing of values and interests among all parties in delivering excellent performance. In addition, the further the partnering process percolates down the supply chain, the greater will be the total achievement.

A good book I faithfully recommend to you.

July 2009

Preface

Relational Contracting, as it applies in construction, is examined in close detail in this book. While acknowledging the fact that as yet there is no agreed definition of the term, four general types of construction procurement management arrangements are nevertheless considered by the authors to embrace the essential ingredients of Relational Contracting. These four comprise project and strategic partnering, project and strategic alliancing, public-private partnerships and joint ventures.

Traditionally, the construction project end product is completely defined in advance of the selection of a main contractor. That contractor agrees under the terms of a formal contract to deliver the defined end product on a date agreed with the client and at an agreed cost. The financial risks are borne by the contractor. In practice, this neat scenario is rarely played out as just described. The contract waters are usually muddied in several ways. Firstly, for instance, during the course of construction the client may want to vary the design, the materials, the time scale, for example, for perfectly good end use or financial reasons. Secondly, risks to the contractor will usually be hedged, to take into account for example risks associated with extreme weather and unforeseeable ground conditions. Agreement on when these conditions actually pertain is not always easy to achieve. Thirdly, in achieving the defined required built quality, the approval by the client of the finished work often becomes a matter of judgement, negotiation and compromise.

For such reasons as these, the relationship between client and contractor is commonly adversarial in nature as agreements concerning client changes and unclear situations are reached and cost effects negotiated.

The problems with the traditional approach have long been discussed within the industry and by academics, and some alternative contract forms have arisen and been in regular use for many decades for larger projects. For example, target cost reimbursable contracts and actual costs plus fixed fee contracts represent attempts by clients to share financial risk. Design and build contracts bring designers and constructors together at the design stage, almost certainly resulting in better value for money. All of these examples tend to promote cooperation between the different parties to a construction project.

Relational Contracting forms, however, are fundamentally different. They have not arisen in order to provide answers to the weaknesses of the traditional approach, or at least not directly. Rather they exist because of the size and complexity of many modern projects, the sophistication of big client bodies, the growth of international markets, the global nature of many companies and the emergence of big companies which are nevertheless quite specialized and expert and must combine their expertise with that of others. In the case of public-private partnerships, in addition to most of the above factors a frequent desire of governments today is to recruit private sector finance and expertise for public sector projects by offering suitable financial incentives to those private providers.

All forms of Relational Contracting however, whatever an ultimate definition might be, and whatever the source of their origins, are directly concerned with the promotion of teamwork among the various parties to a project. Through their research, the authors have identified the following five key elements, common to all successful RC projects, of commitment, trust, cooperation and communication, common goals, and the belief among the parties that arising out of the cooperation, a win-win situation exists for them all.

Construction industry professionals will find in this 'must have' book a comprehensive description of the details of the various forms of Relational Contracts as practiced in the industry, together with discussions of the principles involved. In addition, the very many case studies in the book provide the readers with a good picture of the wide range of project types utilizing RC principles.

Detailed treatment of the considerable amount of academic research on the subject, together with the results of the authors' own research, is worked in throughout the text. The aim of all research, of course, is to illuminate, and generate an understanding of a subject area. The research described in this book has enabled systematic identification of those features which on the one hand are essential to RC success and on the other, merely desirable. The authors have courageously distilled recommendations from the knowledge gained through their intensive research as to those criteria which must be present if an RC application is to be successful.

Professor Michael Anson
Emeritus Professor of Civil Engineering
The Hong Kong Polytechnic University, July 2009

Abbreviations

ACA	Australian Constructors Association
BOO	Build Own Operate
BOT	Build Operate Transfer
BOOT	Build Own Operate Transfer
CII	Construction Industry Institute
CIIA	Construction Industry Institute Australia
CIRC	Construction Industry Review Committee
CPI	Collaborative Process Institute
CSFs	Critical Success Factors
DBFO	Design, Build, Fund and Operate
EU	European Union
GMP	Guaranteed Maximum Price
HKHA	Hong Kong Housing Authority
HKHS	Hong Kong Housing Society
IA	Incentivization Agreement
JV	Joint Venture
KPIs	Key Performance Indicators
KRAs	Key Result Areas
MBA	Master Builders Australia
MTRCL	Mass Transit Railway Corporation Limited
NBCC	National Building and Construction Council
NEC	New Engineering Contract
NOPs	Non-Owner Participants
NPWC	National Public Works Conference
O&M	Operation and Management
PA	Project Alliancing
PFI	Private Finance Initiative
PHCA	Parliament House Construction Authority
PP	Project Partnering
PPP	Public-Private Partnerships
QDMR	Queensland Department of Main Roads
RC	Relational Contracting
SA	Strategic Alliancing
SP	Strategic Partnering
TCC	Target Cost Contracting

Part I
Relational contracting principles

Chapter 1

Fundamentals of relational contracting

Introduction

This chapter begins by depicting the fundamental concepts of relational contracting in general and in the construction industry particularly. Afterwards, it goes into deeper details of the six major types of relational contracting, namely project partnering, strategic partnering, project alliancing, strategic alliancing, public-private partnerships (PPP) and joint venture.

Fundamentals of relational contracting

Relational contracting[1] (RC) is a subject that originally attracted attention in the 1960s (McInnis, 2000). Macneil (1974) advocated that according to RC, a 'contract' can be treated as a 'present promise' of doing something in the future that has dynamic and continuous states of interrelated past, present and future. In other words, contract, which involves present communication of a commitment to a future event, is a projection of exchange into the future (Kumaraswamy et al., 2005). It should be noted that present promise would affect the future by limiting choices that would be available during contract execution. In addition, because of uncertainty and complexity, all future events cannot be easily perceived or quantified. Macneil (1974, 1980) stated that thus contracts should be flexible so as to adjust for future events and effectively address the uncertainties as and when they arise. Macaulay (1963) defined RC as the working relationship amongst the parties who do not always follow the legal mechanism offered by the written contracts, and the parties themselves govern the transactions within mutually acceptable social guidelines. Therefore, the relationship itself develops obligations among the contracting parties.

Rahman and Kumaraswamy (2005) stated that based on an emerging branch of modern contract law, RC provides necessary flexibility to contracts and essential elements of team building. RC dynamically takes into account the events before and after the moments of contract formation. Table 1.1

summarizes various characteristics of classical and modern contract law in relation to construction projects. Eisenberg (2000) stated that RC recognizes that the 'formation of contract is a dynamic, evolving process'. RC emphasizes the ongoing relationships among the contracting parties for the success of a contract and it encourages long-term provisions and mutual future planning, and introduces a degree of flexibility into the contract, by considering a contract to be a relationship among the parties (Macneil, 1974, 1980). RC holds that the world of contract is a world of relations but not a world of discrete transactions. It is a continuous dynamic state, and all segments of 'past, present and future' are interrelated (Macneil, 1974). Goetz and Scott (1981) reckoned that in general, RC principles are frequently seen in the 'best effort' clauses, when parties fail to devise precise performance specifications and well-defined obligations, and thus project 'exchange' into the future and rely on continuous relationships.

Relational contracting in construction

The construction industry has long been suffering from little cooperation, limited trust and ineffective communication, thus resulting in an adversarial relationship between different parties (Moore et al., 1992; Chan et al., 2004b). In fact, construction project teams are composed of different hierarchically and interlinked parties, such as clients/owners, architects, engineers, quantity surveyors, main/general contractors, subcontractors, specialist contractors and suppliers. As a result, complicated relationships exist within project teams and they can adversely affect a project's performance if they are not

Table 1.1 Characteristics of classical and modern contract law (adapted from Rahman and Kumaraswamy, 2005)

Classical contract law	Modern contract law
Rules were usually binary	Rules are often versatile
An overriding preference for objective and standardized rules	Highly flexible in adopting rules that are individualized and even subjective
Principally static	Principally dynamic
Static rules of interpretation that take into account of events only at the moment of contract formation	Dynamic rules of interpretation that consider events before and after the moment of contract formation
Static legal-duty rule	Dynamic modification regime that considers the value of ongoing reciprocity
Static review of liquidated damages provisions	Dynamic review of liquidated damages that takes into account of the actual loss
Only bargains are enforceable	Subjective elements play a critical role in the basic principles of contract interpretation

managed effectively (Walker, 1989). RC has thus been introduced to be an innovative and non-confrontational relationship-based approach to the procurement of construction services in many countries, such as the USA, the UK, Australia and Hong Kong over the last decade (Palaneeswaran et al., 2003; Kumaraswamy et al., 2005; Rahman and Kumaraswamy, 2005). RC is an approach to manage such complex relationships between the players in construction contracts/teams. The foundation of RC is based on recognition of mutual benefits and win-win scenarios through more cooperative relationship between parties (Macneil, 1978; Alsagoff and McDermott, 1994; Jones, 2000; Rowlinson and Cheung, 2004b; Kumaraswamy et al., 2005). RC embraces and underpins different approaches, mainly including partnering, alliancing, joint venture, other collaborative working arrangements and better risk-sharing mechanisms (Macneil, 1978; Alsagoff and McDermott, 1994; Jones, 2000; Rahman and Kumaraswamy, 2002a, 2004; Rowlinson and Cheung, 2004b). As practiced in the construction industry in its myriad forms, the core of RC is to establish the working relationships between the parties through a mutually developed, formal strategy of commitment and communication aimed at win-win situations for all parties (Kumaraswamy et al., 2005). Sanders and Moore (1992) viewed RC as aiming to generate an organizational environment of trust, open communication and employee involvement. Palaneeswaran et al. (2003) viewed that win-win RC approaches such as partnering and alliancing provide vehicles for clients and contractors to drive towards excellence by achieving quality with greater value. Kumaraswamy et al. (2005) stressed that RC is not a 'one-size-fits-all' guaranteed fix, but it is a philosophy that has to be tailored for each situation to which it is applied. Therefore, companies considering RC should evaluate their business objectives, analyse the role of RC in assisting them to achieve those objectives and determine the appropriate style of collaboration to implement. Rahman and Kumaraswamy (2002a) opined that flexible contract conditions should be provided under the broad umbrella of RC principles and this requires transforming traditional relationships towards a shared culture that transcends organizational boundaries (Construction Industry Institute, 1996). Rahman and Kumaraswamy (2004, 2005) also viewed that RC provides the means to sustain ongoing relations in long and complex contracts by adjustment processes of a more thoroughly transaction-specific, continuous and administrative kind. This may or may not include an initial agreement. However, if it does, the need for the contract may be of less importance. Walker and Chau (1999) believed that RC approaches appear useful in achieving the overall objective, which is to reduce the sum of production and transaction costs. In fact, most disputes arise from unclear and/or inappropriate risk allocation (Kumaraswamy, 1997), and risk allocation and management are considered central to contractual and governance structures (Kumaraswamy et al., 2004). Walker and Chau (1999) also stated that RC offers a cost-effective means of encouraging collectively beneficial behaviour when transactions are exposed to opportunism, but a fully contingent contract is too costly to specify. As a matter of fact, given the many potential permutations, all possible contingencies and their likely outcomes cannot be prepared. In addition, they considered that RC is characterized by the subordination of legal requirements and related formal documents to informal agreements in commercial transactions, such as verbal promises or partnering charters. Ling et al. (2006) perceived that RC principles may be mobilized to offer contractual incentives/flexibility, improve relationships among contracting parties, and lubricate any transactional frictions. Cheung (2001) viewed that RC has to allow certain flexibility so as to enable essential adjustments when necessary. Rowlinson and Cheung (2004b) summarized the characteristics of relational contracts and construction contracts in Table 1.2 and they viewed that RC is multilayered and in Australia, it has been identified that there are three levels, namely the Inspector, Engineer and the

Table 1.2 Characteristics of relational contracts and construction contracts (adapted from Rowlinson and Cheung, 2004b)

	Contracting environment	Effectuation	Dispute resolution
Relational contracts	Cooperative but not defensive; Proactive but not reactive	Flexibility and adjustments provisions to cater for unforeseen contingencies	Relational dispute resolution
Construction contracts	Cooperative and mutual trust is the desired static of contracting	Power to issue variations with associated time and cost adjustments	Alternative dispute resolution

Table 1.3 Issues in relationship management at the three levels (reproduced from Rowlinson and Cheung, 2004a with permission for both print and online use from an editor of the Proceedings of the CIBW107/TG23 Symposium on Globalization and Construction in Bangkok)

	Issues
Inspector	• Maintaining quality • Appropriate working methods
Engineer (Individual)	• Quality and claims, yet not empowered to make final decisions on claims or encouraged in the contract to be forthright on quality • Quality of work life • Opportunity to act in an 'old-fashioned' professional manner
Project manager	• Reduction of claims • Timely completion

Project Manager Levels, at which RC needs to operate and that each level has its own issues (Table 1.3).

Manley and Hampson (2000) viewed RC as an umbrella concept, implying an approach to projects that emphasizes teamwork and cooperation. These core elements of relationship management, i.e. teamwork and cooperation, can be applied across the full range of project delivery systems, from build-own-operate-transfer (BOOT) projects to traditional lump sum contracts. The Australian Constructors Association (1999) defined RC as 'a process to establish and manage the relationships between the parties that aim to remove barriers, encourage maximum contribution and allow all parties to achieve success'. Based on this definition, any type of contract could be an RC. In fact, a number of innovative contracting and management models were developed to encourage a more collaborative approach to project delivery. These included: (1) partnering; (2) increasing use of incentives; (3) open-book cost reimbursable performance incentive contracts; and (4) project alliances (Ross, 2003). Table 1.4 shows the comparisons of the definitions of six major types of RC while Table 1.5 compares the similarities and differences of major attributes of six major types of RC.

Major types of relational contracting

As mentioned previously, there are six major types of RC, which encompass (1) project partnering; (2) strategic partnering; (3) project alliancing; (4) strategic alliancing; (5) public-private partnerships; and (6) joint venture.

Project partnering

A number of definitions of partnering have emerged from past studies. Chan et al. (2001) took the view that partnering is a process of building up a moral contract or charter among the project team members, which will tie each party to act in the best interest of both the project and the project team members. Crowley and Karim (1995) conceptually defined partnering as an organization to resolve conflicts, expedite decision-making and increase organizational competence in achieving project goals (Figure 1.1).

The Construction Industry Institute (1991) in the USA and the Construction Industry Board (1997) in the UK conducted some distinguished research into partnering and developed definitions which are widely applied and cited in academic community.

The Construction Industry Institute (USA) defined partnering as:

A long-term commitment between two or more organizations for the purposes of achieving specific business objectives by maximizing the effectiveness of each participant's resources. This requires changing traditional relationships to a shared culture without regard to

Table 1.4 Comparisons of the definitions of six major types of relational contracting

Six major types of RC	Definition
Project partnering (PP)	A long-term commitment between two or more organizations for the purpose of achieving specific business objectives by maximizing the effectiveness of each participant's resources. This requires changing traditional relationships to a shared culture without regard to organizational boundaries. The relationship is based on trust, dedication to common goals and an understanding of each other's individual expectations and values (Construction Industry Institute, 1991).
Strategic partnering (SP)	The major difference between project partnering (relationships established for a single project) and strategic partnering (a long-term commitment beyond a discrete project) is that the former is for a single project (Construction Industry Institute, 1991) but the latter involves at least two projects (Bennett and Jayes, 1998).
Project alliancing (PA)	A cooperative arrangement between two or more organizations that forms part of their overall strategy and contributions to achieving their major goals and objectives for a particular project (Kwok and Hampson, 1996). With alliancing, there is a 'joint' rather than 'shared' commitment. Parties agree on their contribution levels and required profit beforehand and then place these at risk. If one party in the alliance under-performs, then all other alliance partners are at risk of losing their rewards (profit and incentives) and could even share losses according to the agreed project pain-sharing/gain-sharing model (Walker et al., 2000a, 2002).
Strategic alliancing (SA)	The major difference between project alliancing and strategic alliancing is that project alliancing has a defined end, which is most commonly the practical completion date of a project (Peters et al., 2001). However, a strategic alliance usually exists between two companies that extends beyond a specific project (Walker et al., 2000a).
Public private partnerships (PPP)	The collaborations where the public and private sectors both bring their complementary skills to a project, with different levels of involvement and responsibility, for the sake of providing public services (Hong Kong Efficiency Unit, 2003).
Joint venture (JV)	Joint ventures involve two or more legally distinct organizations (the parents), each of which shares in the decision-making activities of the jointly owned entity (Geringer, 1988).

Table 1.5 Similarities and differences of major attributes of six major types of RC

Major Attributes	PP	SP	PA	SA	PPP	JV
Gain-share/Pain share	✕	✕	√	√	√	√
Legal binding	✕	✕	√	√	√	√
Formal contract	✕	✕	√	√	√	√

PP = project partnering   SP = strategic partnering   PA = project alliancing
SA = strategic alliancing   PPP = public-private partnerships   JV = joint venture

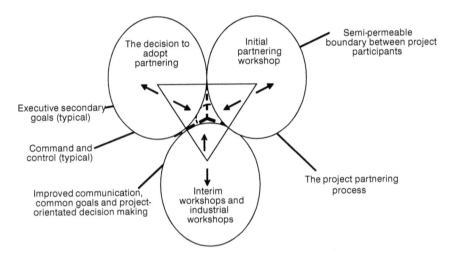

Figure 1.1 Conceptual model of partnering (adapted from Crowley and Karim, 1995)

organizational boundaries. The relationship is based on trust, dedication to common goals, and an understanding of each other's individual expectations and values.

(Construction Industry Institute, 1991)

The Construction Industry Board (UK) defined partnering to be:

A structured management approach to facilitate team working across contractual boundaries...it should not be confused with other good project management practices, or with long-standing relationships, negotiated contracts, or preferred supplier arrangements, all of which lack the structure and objective measures that must support a partnering relationship.

(Construction Industry Board, 1997)

Latham (1994) stated that project partnering provides a channel to introduce participants to partnering philosophy without the need for a long-term commitment. It might benefit future projects or for a longer commitment to a current one (Wilson et al., 1995). Latham (1994) also introduced the process of project partnering, which mainly encompasses the decision to adopt partnering, the initial partnering workshop, the interim workshops and the final workshop (Figure 1.2).

The crucial stages of project partnering explained by Latham were similar to Abudayyeh's (1994) ideas. According to Abudayyeh (1994), it is better for the partnering process to be implemented at the very beginning of the project. The partnering process should be included the expression of interest, first partnering workshop, follow-up workshops and final workshop.

The concept of project partnering was developed by Charles Cowan in the US context in the early 1990s. Manley and Hampson (2000) defined partnering as a commitment between the client and the contractors to actively cooperate to meet separate but complementary objectives. It is associated with the use of a range of tools, encompassing: (1) partnering charters; (2) partnering workshops; (3) team-building exercises; (4) dispute-resolution mechanisms; (5) benchmarking; (6) total quality management; and (7) business process mapping (Bresnen and Marshall, 2000a, b). Walker et al. (2002) stated that project alliancing is different from partnering in that it is more all-embracing in its means for achieving unity of purpose between project teams. It can be seen as occupying the position on a partnering continuum of alliance partners coalescing into a virtual company as illustrated in Figure 1.3.

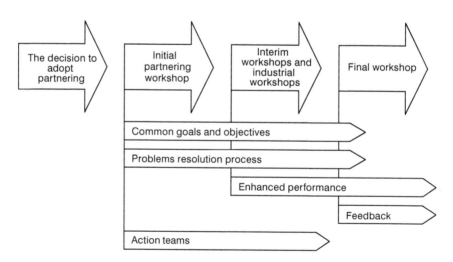

Figure 1.2 The project partnering process (adapted from Latham, 1994)

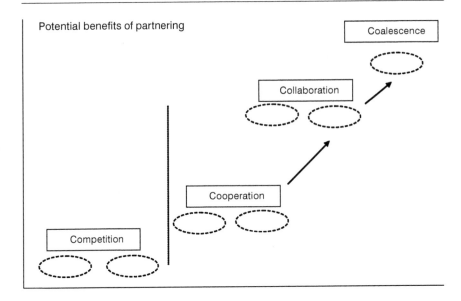

Figure 1.3 Degree of objectives alignment (adapted from Thompson and Sanders, 1998)

The traditional project partnering approach was further modified to 'extended partnering' by the Department of Main Roads in Queensland, Australia (Main Roads Project Delivery System, 2005) in the late 1990s. Main Roads Project Delivery System (2005) stated that extended partnering is a formal process used to facilitate greater team participation and communication outside of the contractual process. It is adopted to develop a cooperative culture between all parties in a contract to achieve excellent project outcomes. It does this by achieving some initial impetus towards achieving harmonious working relationships and shared goals at the beginning of the project, providing suitable problem-solving and team-building skills to the parties to the contract to achieve and maintain these, and providing an agreed structure to assist in maintaining and improving these relationships. Figure 1.4 shows different elements of the extended partnering process.

It is of interest to note that the partnering process and agreements formed in the extended partnering process does not replace the requirements of the contract because there is still a requirement to adhere to contractual notices, timelines and directions. However, the information flow is faster and more comprehensive than that provided for in the contract. Issue resolution durations are also far shorter than provided for in the contract and there is more focus on joint objectives rather than the pursuit of the individual goals of the parties (Main Roads Project Delivery System, 2005).

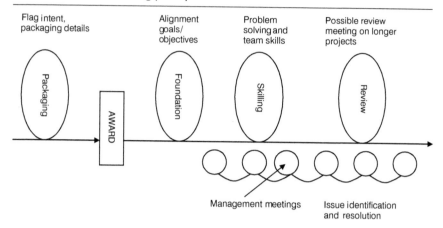

Figure 1.4 A typical 'extended' partnering process (reproduced from Main Roads Project Delivery System, 2005 with permission for both print and online use from the Department of Main Roads, Brisbane, Australia)

In general, typical extended partnering processes as implemented by the Department of Main Roads are as follows (Main Roads Project Delivery System, 2005):

1 Provision of an intent to proceed with the relationship management process either in the invitations for tender or in the contract documents.
2 A foundation workshop generally lasting for 1.5 days conducted early after the award of the contract to align different contracting parties towards common objectives; to develop performance measures for these objectives; to develop a problem resolution process; to establish positive communication channels to enhance relationships; and to develop plans and process for further managing this.
3 A series of relationship management meetings of approximately 2 hours; held at approximately monthly intervals to monitor the common objectives, promote communication, raise and discuss issues and develop plans to resolve them.
4 A skilling workshop is held at about 1 to 1.5 months after the foundation workshop to provide team, communication, motivation, problem-solving and decision-making skills to all project parties. The duration of this workshop is approximately 1.5 days.

Strategic partnering

The major difference between project partnering (relationships established for a single project) and strategic partnering (a long-term commitment beyond a discrete project) is that the former is for a single project (Construction

Industry Institute, 1991; Li et al., 2000; McGeorge and Palmer, 2002; Walker et al., 2002) but the latter involves at least two projects (Construction Industry Institute, 1991; Bennett and Jayes, 1998; Li, et al., 2000; Cheng et al., 2004).

Project alliancing

There are plenty of definitions of alliancing and the scope of alliances is reflected in a range of definitions which are in common currency (McGeorge and Palmer, 2002; Yeung et al., 2007a). These definitions can be very broad, such as 'a relationship between two entities, large or small, domestic or foreign, with shared goals and economic interests' (United States Trade Centre, 1998), or 'organizations with capabilities and needs [that] come together to do business and add value to the other partner, simultaneously working to provide a product which enhances society and the capability of the ultimate client' (Nicholson, 1996). Other authors are more specific, for instance, Kwok and Hampson (1996) defined project alliancing as 'a cooperative arrangement between two or more organizations that forms part of their overall strategy, and contributions to achieving their major goals and objectives for a particular project'. Gerybadze (1995), however, defined project alliancing as follows: the client and associated firms will cooperate for a specific project, but will remain legally independent organizations. Ownership and management of the cooperating firms will not be fully merged although the risk of the project is shared by all participants. The Australian Constructors Association (1999) defined project alliancing as a form of RC involving sound delivery strategies and techniques to optimize project outcomes and deliver optimum commercial benefits for all parties involved. Walker et al. (2000, 2002) analysed that with alliancing, there is a 'joint' rather than 'shared' commitment. Parties consent to their contribution levels and required profit beforehand and then place these at risk. If one party in the alliance under-performs, all other alliance partners are then at risk of losing their rewards (profit and incentives) and could even share losses according to the agreed project pain-sharing/gain-sharing model.

Ross (2003) perceived that a project alliance is when one or more owners form an alliance with one or more service providers (designer, constructor, supplier, etc.) for the purpose of delivering a specific project. The key features of a 'pure' project alliance include:

1 The parties are all responsible for performing the work and generally assume collective ownership of all risks associated with delivery of the project.
2 The owner pays the non-owner participants for their services based on the following '3-limb' 100% open-book compensation model:

- Limb 1: project costs and project-specific overheads reimbursed at cost based on audited actual costs.
- Limb 2: a fee to cover corporate overheads and 'normal' profits.
- Limb 3: an equitable share of the 'pain' or 'gain' depending on how actual project outcomes are compared with the pre-agreed targets which the parties have collectively committed to achieve based on the guiding principle with sharing the pain and sharing the gain between parties.
3 The project is governed by a joint body where all decisions must be unanimous.

The concept of project alliancing was developed on North Sea oil and gas projects at about the same time by British Petroleum (Thompson and Sanders, 1998). Manley and Hampson (2000) pointed out that a key difference between partnering and alliancing is that the former runs alongside standard contracts, having no contractual force in itself, whereas alliancing arrangements are expressed in contractual form. Although alliancing is both a relationship management system and a delivery system, partnering is not a delivery system. In other words, one can have an alliance contract, but there is no partnering contract (it is just a partnering charter). The charter is an agreement signed by all relevant parties showing their interest to cooperate on a construction project. Alliancing is more resource intensive than partnering because of its complicated selection procedures and high continuous information requirements during projects. In its turn, partnering is more resource intensive than conventional approaches because of charter and workshop costs during the initial stage of the project (Manley and Hampson, 2000).

Strategic alliancing

A common definition of strategic alliancing proposed by Love and Gunasekaran (1999) is to build up inter-organizational relationships and to engage in collaborative behaviour for a specific purpose. The major difference between project alliancing and strategic alliancing is that project alliancing has a defined end, which is most commonly the practical completion date of a project. The parties are brought together for a specific project or outcome (Kwok and Hampson, 1996; Manley and Hampson, 2000; Peters et al., 2001; Walker et al., 2002; Rowlinson and Cheung, 2004b; Hauck et al., 2004). Nevertheless, a strategic alliance often exists between two companies that extends beyond a specific project (Hampson and Kwok, 1997; Walker, et al., 2000; Peters et al., 2001; Rowlinson and Cheung, 2004b; Hauck et al., 2004) and the parties would expect mutually beneficial continuous business (Peters et al., 2001).

Public-private partnerships

Public-private partnerships (PPP) or private finance initiatives (PFI) were originated in the UK in the late 1980s and early 1990s (Asian Development Bank, 2007). Since its inception, PFI has become the preferred method to procure public infrastructure projects and services. Nowadays, PPP has been widely used to procure public infrastructure projects in developed nations and in sectors beyond public infrastructure. Asian Development Bank (2007) stated that there are a number of forms and structures of PPP, such as build-operate-transfer (BOT), build-own-operate (BOO), leasing, operation and management (O&M), equity joint venture, and cooperative joint venture, and there is neither a standard framework of PPP practices nor a fixed definition of PPP. Hong Kong Efficiency Unit (2003) defined PPP as 'the collaborations between public and private sectors to bring their complementary skills to a project, with various levels of involvement and responsibility, for the sake of providing public services'.

The core of PPP is that it is a market-driven approach for government procurement of assets and/or services when compared with the traditional approach where these have been fully funded and undertaken by the government. In fact, PPP is always beneficial to improved quality of public assets and services, cost savings and relief of public finance and increased commercial opportunities between the government and private sector participants (Asian Development Bank, 2007).

A typical PPP project arrangement can be divided into several stages, encompassing: (1) pre-qualification; (2) request for proposals; (3) best and final offer; (4) negotiations and financial closure. Parties involved in a PPP project usually encompass government departments that are charging the PPP procurement; the private sector participants who may form a special project vehicle responsible for delivery of the PPP project (Asian Development Bank, 2007). As a matter of fact, different parties involved in PPP always have conflicting objectives and interests. Therefore, for the sake of implementing PPP smoothly, there is a great deal of negotiations amongst all parties involved in PPP projects to ensure that rights and responsibilities are clearly defined, risks are equitably allocated, and mitigation measures exist in case of default (Asian Development Bank, 2007).

Joint venture

Joint venture (JV) involves two or more legally individual organizations (the parents), each of which shares in the decision-making activities of the jointly owned entity (Geringer, 1988). Tomlinson (1970) defined JV as an arrangement where there is commitment of funds, facilities and services by two or more legally separated interests to an enterprise for their mutual benefit for a long period of time. An international joint venture (IJV) is regarded as at least one parent being headquartered outside the venture's country of operation, or if the JV has a significant level of operations in more than one nation (Geringer and Hebert, 1989). Traditionally, JVs are often used to exploit peripheral markets or technologies, and their activities are typically of marginal importance to the maintenance of a parent firm's competitive advantage. Nevertheless, Harrigan (1987) stated that JVs have been increasingly perceived as critical elements of an organization's business unit network and as strategic weapons for competing within a firm's core markets and technologies.

Carrillo (1996) viewed that JV is mainly a tool of convenience allowing the parties involved to exploit each other's strengths for a limited period of time and for a specific cause. Having interviewed eight UK contractors, Carrillo (1996) highlighted the perceived advantages and disadvantages of JV as discussed in Table 1.6.

In fact, there may be inherent difficulties with the use of JV to transfer technology. Ofori (1991) stated that foreign companies may be unwilling to impart their skills, and local partners may be unable to benefit from the knowledge and skills on offer. The World Bank (1986) also perceived that JVs are only effective if a developing country has at least some managerial skills. Walker and Johannes (2003) stated that alliances and JVs can provide a tool for creating agility in response to the diversity in skills, work culture

Table 1.6 Joint venture advantages and disadvantages (reproduced from Carrillo, 1996 with permission for both print and online use from Taylor & Francis Journals (UK))

	Advantages	Disadvantages
Partner from developed country	• Spreading of risks • Sharing of fixed cost • Pooling of knowledge and resources • Reduces cultural problems • As a front • Political influence • Local purchasing muscle	• Ability to reduce potential competition • Unstable agreements for a limited period • Reduces competitive advantage • Local partner may continually want more • Additional risk • Loss of control • Diluted earnings • Possible conflict
Partner from developing country	• Balance sheet substance • Improved track record • Sharing of risks • Opportunity to work on large projects • Staff acquires know-how • Improved purchasing power	• May be considered as necessity only • Loss of control • Work goes to overseas company • Possible dilution of 'single point of responsibility' principle

and business practices that characterizes customers. In fact, JVs are common in the construction industry especially in forming consortia to undertake a BOT/BOO/BOOT project. The input stake of JV partner differs. For instance, an entrepreneurial stake identifies customer need, develops the project concept or locates the financing entities, develops the legal structure for a project ownership group and negotiates the operating concession (Walker and Johannes, 2003). Walker and Johannes (2003) also mentioned that there are a number of motivations for forming construction JVs in large-scale construction projects – in addition to the more obvious one of providing sufficient financial strength to participate in very capital-hungry infrastructure projects – and one of them is to reduce risk exposure to clients that may have cash-flow or other financial problems encompassing a capacity to pay tolls on road, bridge or tunnel projects. However, significant problems may arise with political and cultural differences stemming from one JV partner not understanding the political and historical pressures influencing their partners.

Chapter summary

This introductory chapter first describes the fundamentals of RC both in general and particularly in construction. Then, it goes into the details of the six major types of RC, including (1) project partnering; (2) strategic partnering; (3) project alliancing; (4) strategic alliancing; (5) public-private partnerships; and (6) joint venture. In Chapter Two, the global development of RC will be first narrated, followed by depicting the RC developments in the USA, the UK, Australia and Hong Kong.

Chapter 2

Relational contracting development in major countries/cities

Global development of relational contracting

The global construction industry is commonly accepted as a very competitive and risky business because there always exist a number of problems, such as lack of cooperation, insufficient trust and ineffective communication often inducing an adversarial relationship among all key project stakeholders. The confrontational relationship is likely to lead to poor project performance in terms of time, cost and quality (Moore et al., 1992). In the late 1980s, professional bodies began to recognize that if the construction industry was to compete for investment funds, particularly worldwide, both the methodology and the public image of the construction industry would have to be improved. Different inquiries were conducted into the practices and productivity of the building and construction industry, as a result of which some important reports were published to help improve the industry, like Ireland (1988); Smith (1988); Parliament House Construction Authority (PHCA) (1990); National Public Works Conference (NPWC)/National Building and Construction Council (NBCC) (1990); Gyles (1992); Latham (1994); Egan (1998); and Construction Industry Review Committee (CIRC) (2001). Over the past two decades, RC has been acknowledged by many practitioners and academics as an innovative approach to the procurement of construction services in the construction industry.

Partnering, the first and the most common type of RC all over the world, was initially developed by the US Army Corps of Engineers as a project delivery strategy (Cowan et al., 1992). Agencies in the US developed partnering as a formal management process designed to facilitate better understanding and closer collaboration between contracting parties while still working under a traditional form of contract (Department of Treasury and Finance, 2006). Afterwards, it became an established approach to contracting in the USA, the UK, Australia and Hong Kong since the 1990s (Bresnen and Marshall, 2000a, b). In fact, the paces of RC development are different in different places. Although partnering as a kind of RC was originally developed in the USA, its developments in the UK, Australia and Hong Kong are much more rapid. The New Engineering Contract (NEC) first published in the UK in 1993 was depicted as a modern day family of standard contracts that truly embraces the concept of partnership and encourages employers, designers, contractors and project managers to work together through both a powerful management tool and a legal framework to facilitate all aspects of the creation of construction projects.

(Department of Treasury and Finance, 2006)

In fact, Bennett and Jayes (1998) stated that partnering was developed in the UK in the 1980s, which was then further developed to be so-called 'second generation' and 'third generation' in the 1990s. Project alliancing was first used in the UK in the early 1990s to deliver step-change improvements in the delivery of complex offshore oil and gas projects.

The Australian Constructors Association (ACA) (1999) viewed that RC has been successfully implemented by leading Australian construction companies on a diverse range of projects. Project-specific partnering was first promoted by Charles Cowan in the 1980s through his work with the US Army Corps of Engineers and the Arizona Department of Transportation (Ross, 2003). The traditional project-specific partnering approach was further modified to 'extended partnering' as adopted by the Department of Main Roads in Queensland in the late 1990s (Main Roads Project Delivery System, 2005).

Nearly at the same time, project partnering was further developed into what is known as project alliancing.

Project alliancing is different from project partnering in that it is more all-embracing in its means for achieving unity of purpose between project teams (Walker et al., 2002). Under partnering, one team may 'sink or swim' without necessarily affecting the business position of other teams. However, for alliancing, there is a joint, rather than shared, commitment. If one party in the alliance under-performs then all other alliance partners are at risk of losing their rewards and could even share losses according to the agreed project gain-sharing/pain-sharing model

(Walker et al., 2000).

In Hong Kong, partnering was first adopted in a hospital project in 1994 and it was then increasingly applied in the public, private and infrastructure sectors. Not only the traditional 'structured', but also the 'unstructured' approach was adopted in many construction projects. It should be noted that partnering in Hong Kong was always implemented successfully together with target cost contracting (TCC) principles, such as guaranteed maximum price (GMP) and incentivization agreement (IA).

It is widely accepted that public-private partnerships (PPP) offers a long-term and sustainable method to improve social infrastructure, enhance the value of public assets, and make better use of taxpayers' money (Li and Akintoye, 2003). In fact, the concept of PPP in the USA and Europe has existed for more than a century, but has become more primary over recent decades in local economic development (Keating, 1998). Li and Akintoye (2003) stated that partnerships come in all sizes and types which make it hard to group them in a consistent fashion. The most vital PPPs since the 1990s have been in the sectors of education, health and transportation. Asian Development Bank (2007) stated that there are a number of forms and structures of PPP, such as build-operate-transfer (BOT), build-own-operate (BOO), leasing, operation and management (O&M), equity joint venture and cooperative joint venture, and there is neither standard framework of PPP practices nor a fixed definition of PPP. As a matter of fact, the pace and magnitude of joint venture development are different in different nations. In the USA, an increased interest in the privatization of government-produced services has been observed since the 1990s, especially at the country and municipal levels (Li and Akintoye, 2003). In the UK, PPP is a key element in the government's strategy for delivering modern, high-quality services and promoting the UK's competitiveness (HM Treasury, 2000). PPP covers a range of business structures and partnership arrangements, from the private finance initiative (PFI) to joint ventures and concessions, outsourcing, and the sales of equity stakes in state-owned business. PPP is also becoming popular in Australia. Li and Akintoye (2003) stated that New South Wales, as well as other states of Australia, has faced increasing demands for infrastructure of all types. It has been difficult to meet these demands because of budgetary constraints. In China, a number of large-scale infrastructure projects have adopted a PPP approach through the BOT schemes to deliver their projects (Li and Akintoye, 2003).

Similar to PPP, Ozorhon et al. (2008) stated that joint venture has become popular because of its importance as a strategic alternative in global competition. Hu and Chen (1996) stated that international joint venture has become a major form of business and is widely used by multinational corporations. And the formation of international joint ventures creates corporate synergism. Local partners, especially those from developing nations, would benefit from the technological know-how, management skills and capital brought in by their foreign partners. On the other hand, multinational corporations can use local partners' knowledge and connections in the host nations to reduce risks and increase revenue. Contractor and Lorange (1988) viewed that economies of scale can be achieved through trust and cooperation among the partners. Therefore, international joint ventures allow partners to share information, resources, markets and risks, hence creating synergistic effects. Since the early 1990s, there has been a sharp increase in the number of Sino-foreign joint venture construction projects in the People's Republic of China (Gale and Luo, 2004; Chan and Suen, 2005). Chan and Suen (2005) stated that although the Chinese government has adopted new measures to boost foreign involvements, it is still problematic for many foreign contractors to enter into the construction market and one of the potential problems concerns construction disputes.

Geringer and Hebert (1989) stated that JV is often traditionally used to exploit peripheral markets or technologies, and their activities are typically of marginal importance to the maintenance of a parent firm's competitive advantage. Nevertheless, Harrigan (1987) stated that JVs have been increasingly perceived as critical elements of an organization's business unit network and as strategic weapons for competing within a firm's core markets and technologies. Hu and Chen (1996) opined that although joint venture can help improve efficiency, it has also created problems for the partners because of different goals, values and cultures. Beamish (1985) viewed that the developing countries are considered a more complicated and difficult environment in which to manage joint ventures than developed nations. In addition, the developing nations are characterized by a higher instability rate and greater managerial dissatisfaction than that of developed nations. Carrillo (1996) viewed that JV is mainly a tool of convenience allowing the parties involved to exploit each other's strengths for a limited period of time and for a specific cause. In fact, there may be inherent difficulties in the use of JV to transfer technology. Ofori (1991) stated that foreign companies may be unwilling to impart their skills, and local partners may be unable to benefit from the knowledge and skills on offer. The World Bank (1986) also perceived that JVs are only effective if a developing country has at least some managerial skills.

Walker and Johannes (2003) stated that alliances and JVs can provide a tool for creating agility in response to the diversity in skills, work culture and business practices that characterizes customers. In fact, JVs are common in the construction industry especially in forming consortia to undertake a BOT/BOO/BOOT project. The input stake of JV partner differs. For instance, an entrepreneurial stake identifies customer needs, develops the project concept or locates the financing entities, develops the legal structure for a project ownership group and negotiates the operating concession (Walker and Johannes, 2003). Walker and Johannes (2003) also mentioned that

> there are a number of motivations for forming construction JVs in large-scale construction projects in addition to the more obvious one of providing sufficient financial strength to participate in very capital-hungry infrastructure projects, and one of them is to reduce risk exposure to clients that may have cash-flow or other financial problems encompassing a capacity to pay tolls on road, bridge or tunnel projects. However, significant problems may arise with political and cultural differences stemming from one JV partner not understanding the political and historical pressures influencing their partners.

Large-scale construction infrastructure projects in Hong Kong have tended to be delivered using JV of global and local construction organizations. The major reason why individual, usually very large, construction companies do so is to offer a customer-focused service package that meets that customer's needs. Supporting reasons include bridging knowledge and expertise gaps, sharing risk and exploring opportunities to add value to the JV organizations through collaboration (Walker and Johannes, 2003).

Relational contracting development in the USA

Partnering, as an initial and major type of RC, appears to be a management tool that encourages greater integration of the project team and creates competitive advantages to all that participate in the project. It is believed that the concept originates from the USA from the early 1980s where team building, cooperation and equality, rather than the single-sided relationship of adversaries to a project, were encouraged (Naoum, 2003). In its original form in the USA, partnering was clearly intended to be a post-procurement selection process aimed at minimizing risk and maximizing profit (Edelman et al., 1991). Gerard (1995) mentioned that in the USA, partnering was first developed by the Portland, OR district of the US Army Corps of Engineers. Weston and Gibson (1993) conducted a study of US Army Corps of Engineers' construction projects and found that 31 out of 37 domestic districts used partnering and the survey highlighted 19 partnering projects and used data available for 16 of them which represented approximately 85 per cent of projects undertaken using partnering at that time. By using criteria for success of cost change, change order cost, claim cost, value engineering savings and duration change, a comparison of these was made with 28 non-partnering projects using t-tests. The results indicated that the partnering projects had achieved much better project performance in terms of all these criteria. Further evidence, based on the US Construction Industry Institute (CII) study of partnering experience gathered from 21 partnering relationships involving more than 30 owners, showed an overwhelmingly positive experience with partnering and validated the choice to partner (Construction Industry Institute, 1991) and the results are illustrated in Table 2.1.

Since then, the real growth of partnering in the USA has been on a project-by-project basis mainly within the public works sector (Loraine, 1994). Gerard (1995) also viewed that partnering has proved to be a very effective process in the US and it was increasingly used in both the public and private sectors. Thompson and Sanders (1998) stated that the potential benefits of partnering are illustrated in terms of the extent that project objects are aligned through partnering, i.e. from competition to cooperation, then to collaboration and coalescence.

The concept of PPP in the USA has existed for more than a century, but has become more prominent over recent decades in local economic development (Keating, 1998). In fact, there has been an increased interest in the privatization of government-produced services since 1990s, particularly

Table 2.1 CII USA study results (adapted from Thompson and Sanders, 1998)

Relationship	Characteristics
Competition (low objectives alignment)	1. No common objectives: they may actually conflict with each other 2. Success occurrence at the expense of others (win-lose mentality) 3. Short-term focus 4. No common project measures between organizations 5. Competitive relationship maintained by coercive environment 6. Little or no on-going improvement 7. Single point of contact between organizations 8. Little trust, with no shared risk; mainly a defensive position
Cooperation (low to medium objectives alignment)	1. Common goals that are project specific 2. Improved interpersonal relationships 3. Team members who are likely to be involved in projects outside the partnering relationship 4. Partnership measures that may or may not be similar to organizational measures used on other projects 5. Multiple points of contact 6. Insufficient trust and shared risk: guarded information sharing
Collaborative (medium to high objectives alignment)	1. Long-term focus on fulfilling the strategic goals of relevant parties 2. Multi-project agreement: long-term relationships without guaranteed workload 3. Common measurement system for the projects and the relationship 4. Improved processes and reduced duplication 5. Relationship-specific measures tied to team incentives 6. Shared authority 7. Openness, honesty and increased risk sharing
Coalescing (high objectives alignment)	1. One common performance measurement system 2. Cooperative relationships supported by collaborative experiences and activities 3. Cultures integrated and directed to fit the application 4. Transparent interface 5. Implicit trust and shared risk

at the country and municipal levels (Li and Akintoye, 2003). And among the services that are historically most commonly contracted out in the USA are solid waste disposal, street construction, management and operation of facilities, building repair, ambulance services, vehicle repair and maintenance, architectural and engineering services, legal counsel, public schools, welfare-to-work programmes, and inner city redevelopment (Florestano and Gordon, 1980; Hirsch and Osborne, 2000; Li and Akintoye, 2003). Martin (1996) also noted that hundreds of thousands of housing units have developed as a result of PPP in the USA. And from a centralized accounting and maintenance system, Florida has structured a portfolio of public-housing properties that are individually managed by private companies as market-rate housing, while retaining the social responsibilities of the public sector.

Relational contracting development in the UK

During the last decade, many reports and commentators (Sir Michael Latham's Report in 1994; the Construction Best Practice Programme in 1998; and Sir John Egan in 1998) have indicated that there is a need for a change in the fundamental culture of the construction team. Perhaps Latham's Report (1994) has proved to be the most significant milestone, in that it indicates for the first time that the public sector should change procedures and methods to incorporate the concept of 'partnering' used so successfully in the USA and Australia. Whilst Latham identified the major problems of the construction team and made recommendations, there remained a critical mass of large, private-sector clients that remained unconvinced that yet another report would lead to change. This culminated in the New Labour government of 1997 teaming with the private sector and the publication of Egan's Report in 1998. Egan's Report represents the views and findings of major clients such as the British Airports Authority, Railtrack, Tesco, National Westminster Bank PLC (NatWest), other major supermarket chains and similar organizations. Egan, representing predominantly private sector interests, accepts, perhaps surprisingly, that the lowest price should not be the only criterion used for selection and that other factors must be considered (Naoum, 2003). In fact, one of the key features of Egan's Report is the recommendation that clients and the construction industry rely less on competitive tendering and formal construction contracts and move to a 'supply chain' system of construction production more commonly found in the manufacturing industry. These features are consistent with the ethos of the partnering process, which tries to steer parties towards trust and cooperation to reduce formal contractual and inevitable adversarial positions, and at the same time eliminate price as the major criterion of selection. After the recommendations of Latham's and Egan's Reports, partnering has steadily gained popularity from the early 1990s and the British Government also supported radical changes in the way that the construction industry performs and provides services to customers, particularly those in the public sector. Client organizations including those such as the British Airports Authority, Railtrack, banks and supermarket chains have all carried out projects using the partnering process (Cook and Hancher, 1990; Hellard, 1996a). In general, all of them reported that results are very positive with large savings in terms of cost and time claimed (Bennett and Jayes, 1998). Loraine (1994) opined that the application of partnering to the UK public works sector was affected by the European Union (EU) public procurement regulations. Bennett and Jayes (1998) viewed that partnering has witnessed three generations of development, each of which are different in concept and design and therefore also different in terms of definition. A study by Walter (1998) revealed how important partnering has become to UK construction companies. Walter (1998) further reported that many firms expected to see partnering account for over two-thirds of their workloads, and companies encompassing AMEC, Amey Construction, Bovis, HBG Construction and Alfred McAlpine Construction all stated that the majority of their current contracts were partnered (Matthews, 1999).

A major UK contractor realized in 1990 that the construction industry could no longer continue to employ the traditional adversarial approach to working with subcontractors and the main contractor believed that in order to work more productively it had to work more closely with its subcontractors and develop closer working relationships (Matthews, 1999). For this, the semi-project-partnering approach has been developed. Although the approach is primarily applicable to the main contractor-subcontractor relationship, the methods used in the semi-project-partnering approach can be adopted throughout the supply chain. The approach was termed 'semi-project-partnering' because it was understood by the main contractor that 'true' partnering would be based on negotiation rather than competition. In fact, the semi-project-partnering approach employs limited competition using the fundamental principles of project partnering. This approach has three main phases: (1) procurement set up: package and company

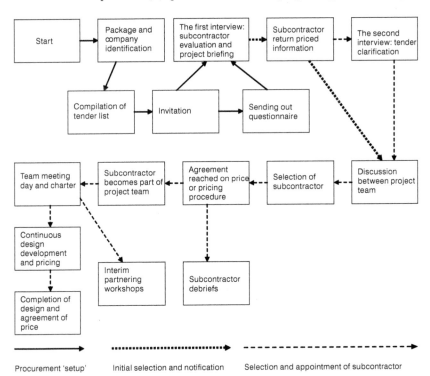

Figure 2.1 Implemented semi-project-partnering approach (adapted from Matthews, 1999)

identification; (2) initial selection and notification: first subcontractor interview and second subcontractor interview; and (3) selection and appointment of subcontractor: third interview and tender clarification, subcontractor selection, project day and pricing (Matthews, 1999). Figure 2.1 shows diagrammatically the sequence of events within the implemented semi-project-partnering approach.

However, Mason (2007) stated that specialist contractors hold a negative view of partnering with a limited take-up due in part to an exclusion from participation. As a matter of fact, many of the industry's main contractors have already taken steps towards applying the principles of partnering to their suppliers. Nevertheless, partnering in the construction industry is a concept that has by no means been universally accepted; this is thought to be particularly so, at and below tier two in the supply chain (Figure 2.2). Bennett and Jayes (1998) stated that partnering was first developed in the UK in the 1980s, which was then further developed as so-called 'second generation' and 'third generation' in the 1990s. And project alliancing was first used in the UK in the early 1990s to deliver step-change improvements in the delivery of complex offshore oil and gas projects.

PPP is an important element in the government's strategy to deliver modern and high-quality services, and to promote the UK's competitiveness (HM Treasury, 2000). PPP covers a wide range of business structures and

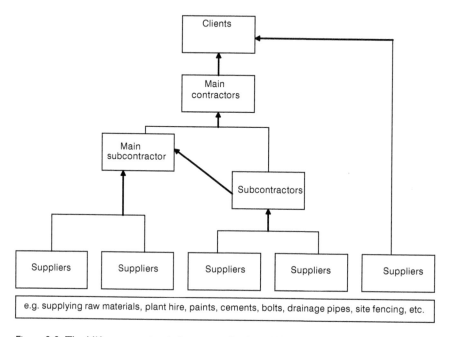

Figure 2.2 The UK construction industry supply chain (Reproduced from Beach et al., 2005, with permission for both print and online use from Elsevier)

partnership arrangements, ranging from PFI to joint ventures and concessions, outsourcing, and the sales of equity stakes in state-owned business (Li and Akintoye, 2003). PFI is the principal model of PPP in the National Health Service (NHS) of the UK. NHS (1999) stated that PPP in the NHS is not only about the financing of capital investments, but also about exploring the full range of private sector management, commercial and creative skills. Li and Akintoye (2003) stated that major PFI schemes are mostly design, build, fund and operate (DBFO), whereby the NHS makes annual payments for the use of privately owned facilities over a concession period of 25 to 40 years. On the other hand, school PPP projects involve buying asset-related services for schools from the private sector. PPP projects usually involve using assets and the contractors provide and operate these assets, and are always able to generate income from them.

Relational contracting development in Australia

Master Builders Australia (MBA) pioneered the partnering concept in Australia in 1992 by bringing Charles Cowan to Australia in September 1992 and again in 1993 and 1994 to conduct a series of seminars and meet industry leaders. The result is that a large number of projects have utilized the partnering concept in the construction and other industries (Wilson et al., 1995). In fact, partnering was extensively used in many civil and building projects around Australia. In the late 1990s, the Australian Constructors Association (ACA) (1999) developed and promoted the concept of RC, which was defined as 'a process to establish and manage the relationships between the parties that aim to remove barriers, encourage maximum contribution and allow all parties to achieve success'. Various states introduced guidelines and codes of practice to improve the standard of administration on government contracts and legislation, providing more protection to those at the lower end of the contracting chain. Table 2.2 shows that there are three types of partnering as defined by the Construction Industry Institute Australia (1996).

The traditional partnering approach was further modified to 'extended partnering' by the Department of Main Roads in Queensland (Main Roads Project Delivery System, 2005). A further development is to take project partnering a step further into what is known as project alliancing. Different from project partnering, project alliancing is more all-embracing in its means for achieving common goals between project teams (Walker et al., 2002). Under partnering arrangements, one project team may have poor project performance without necessarily affecting the business position of other teams. In contrast, there is a joint, rather than shared, commitment under alliancing arrangements. If one project team in the alliance performs badly, all other alliance teams are at risk of losing their rewards and could even share losses according to the mutually agreed project pain-sharing/

Table 2.2 Australian partnering forms (adapted from Construction Industry Institute (Australia), 1996)

Partnering type	Experimental partnering	Packaged partnering	Committed partnering
Partnering description	• Partnering charter, workshop and small number of interim meetings. • Always first partnering experience • Minimally resourced.	• Offered as part of a contractor's tender or imposed upon the contractor after accepting the tender. • Usually only involves the client and the contractor. • This model is adopted very successfully as a marketing tool.	• Often developed as a result of first, unsuccessful experience. • Incorporates as many stakeholders as possible in a tight, well-facilitated dispute-resolution mechanism. • Well resourced
Partnering outcome	• Often unsuccessful, generally due to a lack of clear understanding, commitment and structure.	• Problems may arise due to a lack of commitment and understanding of each stakeholder's objective. • A client/contractor relationship perceived to be cooperative at the outside of a project may not essentially last for the duration of the contract.	• Problems may arise from lack of commitment and understanding of each stakeholder's objective. • A client/contractor relationship perceived to be cooperative at the outside of a project may not essentially last for the duration of the contract.

gain-sharing model (Walker et al., 2000). Historically, project alliances were used to deliver some major oil and gas projects in Western Australia in the early 1990s (Ross, 2003) and Sydney Water used a project alliance to deliver the Northside Storage Tunnel project in the late 1990s. Since then, project alliancing has been used by different states and local governments and by the Commonwealth to deliver a wide range of engineering infrastructures, covering road and rail transport, water supply, storage and treatment, solid and liquid waste, communications, channel and port facilities, defence materials, and other sectors (Department of Treasury and Finance, 2006). The Australian National Museum project provides a useful illustration of project alliancing in Canberra (Walker et al., 2002).

PPP is popular in Australia. Some states of Australia, especially New South Wales (NSW), have encountered increasing demands for infrastructure of all types. As a matter of fact, meeting these demands is difficult due to budgetary restraints (Li and Akintoye, 2003). Raneberg (1994) depicted four key reform objectives of the NSW Government as the optimal allocation of scarce public sector resources, efficiency, better service and accountability for performance. A number of market-oriented initiatives have been employed to secure these objectives, which include measures ranging from wholesale privatization to contracting out in-house service needs, and private sector participation in infrastructure projects.

Relational contracting development in Hong Kong

The earliest formal partnering arrangements recorded within the Hong Kong construction industry were exclusively applied to hospital projects commencing in 1994 (Skues, 1996). The two pioneering proponents were the Hospital Authority and Hsin Chong Construction Co. Ltd., a leading Hong Kong-based multidisciplinary contractor. The two design and build hospitals being managed by the Hospital Authority have embarked on partnering. The first of these projects, the North District Hospital located in the Sheung Shui District, introduced partnering through the initiative of the employer (Skues, 1996). The initial partnering workshop was conducted after tender-out but before the contract award. The second project, the Tseung Kwan O Hospital, included a partnering provision in the contract, and thus the partnering workshop was launched by the contractor, Hip Hing-Laing Joint Venture. Leighton Contractors (Asia) Ltd., a prominent Australian-based contractor, has successfully adopted partnering for the contracts of the Haven of Hope Hospital in 1995 and the United Christian Hospital in 1997 (Chan et al., 2002a). Since then, more than 100 projects have adopted partnering arrangements, following remarkable successes in other countries, such as the USA, the UK and Australia.

In fact, the application of partnering principles has not been limited to hospital projects. The quasi-government mass transportation service providers, Kowloon-Canton Railway Corporation (KCRC) and Mass Transit Railway Corporation Ltd. (MTRCL), have introduced partnering for their development projects such as the West Rail and the Tseung Kwan O Railway Extension (Bayliss, 2002). Since the early 2000s, the MTRCL has achieved outstanding partnering performance because of simultaneously implementing incentivization agreements (IA) and target cost contracts (TCC) (Chan et al., 2008). Moreover, the focus on reducing construction disputes via partnering has been placed in the public sector. Apart from the infrastructure developments, the Hong Kong Housing Authority (HKHA) and the Hong Kong Housing Society (HKHS) are also actively nurturing a partnering culture in public and semi-public sector residential developments (Chan et al., 2002a).

The report of the Construction Industry Review Committee (CIRC) of the Hong Kong SAR Government published in 2001 also recommends the wider adoption of a partnering arrangement in local construction (CIRC, 2001). Partnering creates a better environment to all project participants to work as a team to achieve shared project objectives through collaboration rather than in competition with each other. The CIRC Report (2001) develops the foundation for implementing project partnering in Hong Kong. It also attracts considerable attention of the construction industry in adopting project partnering because of a growing quest for substantial improvements in project performance. With the pioneering support from the local public bodies and government authorities, many private firms are willing to try out partnering approach in their projects. For example, some leading private property developers (e.g. Hongkong Land Ltd. and Swire Properties Ltd.) are adopting the partnering concepts in their prestigious building development projects to reap the perceived advantages. It should be noted that the implementation of partnering, together with guaranteed maximum price (GMP) contractual arrangement, are extremely successful for some building projects. On the other hand, a number of large-scale infrastructure projects have adopted PPP and JV approaches to deliver their projects (Li and Akintoye, 2003; Walker and Johannes, 2003). A main reason why individual, usually very large, construction companies do so is to offer a customer-focused service package that meets that customer's needs, and the most important supporting reasons include bridging knowledge and expertise gaps, sharing risk and exploring opportunities to add value to the JV organizations through collaboration (Walker and Johannes, 2003).

Chapter summary

This chapter follows the evolution of global development of RC. Then, it goes into more detail of RC development in the USA, the UK, Australia and Hong Kong. It has been found that the paces of the development of RC are quite different amongst different regions. Although partnering as a kind of

RC originated in the USA, its development in the UK and Australia are much more rapid. Partnering has experienced the second and third generations in the UK in the 1990s and project partnering was further modified to extended partnering and then project alliancing in Australia in the late 1990s. Following Australia's development, partnering success in Hong Kong is characterized by adopting target cost contracts such as IA and GMP contracts. In addition to these, PPP and JV have been found to be widely applied to procure large-scale construction projects worldwide. And major reasons include knowledge bridging, sharing risk and exploring opportunities to add value to the partnered organizations through collaboration. In Chapter Three, the definition of RC in construction will be analysed using German philosopher Ludwig Wittgenstein's family-resemblance concept.

Chapter 3

The definition of relational contracting in construction as a Wittgenstein family-resemblance concept

Introduction

An increasing trend of the adoption of RC in construction has received much attention from academics and industrial practitioners in the construction industry since the 1990s (Rahman and Kumaraswamy, 2002, 2004; Rowlinson and Cheung, 2004b). Although more and more academic papers have discussed the characteristics of RC, there is still no consensus on the precise and comprehensive meaning of the concept because various researchers view it differently and the paces of RC development in different countries are quite different. In fact, RC can be characterized as a complicated concept where it has been difficult to reach an agreement on a standard type of definition. An explanation for the increasing number of RC definitions is that the concept is yet to mature. If this is true, a comprehensive and conclusive definition of RC, which states the necessary and sufficient conditions, will finally arise. Nevertheless, the reality is just the opposite. It seems that the first step to clearly understanding the concept of RC in construction is probably to realize that such a definition does not exist for this vague and versatile concept.

However, there is still a need for a common concept of RC because discussions will be at cross-purposes and ineffective if there is not any mutual starting point. The aim of this chapter is to present the innovative and useful sunflower model to define RC in construction, which is developed by using German philosopher Ludwig Wittgenstein's family-resemblance concept. The same methodology is applied to define construction partnering (Nyström, 2005) and construction alliancing (Yeung et al., 2007a), which are two major forms of RC in construction.

Previous related research works

To illustrate the usefulness of the same research methodologies to define complicated, vague and multifaceted concepts, previous research works on defining construction partnering (Nyström, 2005) and construction alliancing (Yeung et al., 2007a) are summarized in this section as illustrative examples.

Nyström (2005) conducted a comprehensive literature review on partnering and made two significant contributions to the debate about the definition of construction partnering. The first was a distinction among general prerequisites, components and ultimate goals when partnering is discussed. By conducting a comprehensive literature review, he stated that the focus ought to be on the components so as to grasp what is specific about partnering. The second contribution was to apply Wittgenstein's idea of family-resemblance to the concept of partnering. It was concluded, from the literature review, that there are two necessary components in partnering – trust and mutual understanding – and that a number of various components can be added to form a specific variant of partnering. Figure 3.1 shows the partnering sunflower model containing all the key elements of partnering while Figure 3.2 indicates the applied partnering sunflower model with two case studies. This provides a new method to define the vague and multifaceted concept of partnering in both a flexible and structured way.

Yeung et al. (2007a) observed that there has been increasing interest in the concept of construction alliancing arising from the late 1990s. They stated that although project partnering is a widely understood concept, the same is not true for construction alliancing. By using Nyström's similar approaches to define construction partnering, which include literature review, content analysis and Wittgenstein's family-resemblance concept, the authors focused on alliancing and family-resemblance, and made two contributions to the concept of alliancing in construction. The first one was to distinguish among general prerequisites, hard (contractual) and soft (relationship-

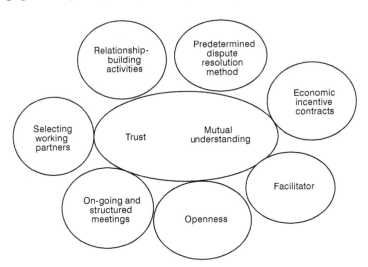

Figure 3.1 Partnering sunflower model containing all the key elements of partnering (reproduced from Nyström, 2005 with permission for both print and online use from Taylor & Francis Journals (UK))

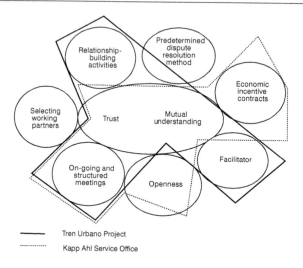

Tren Urbano Project
Kapp Ahl Service Office

Figure 3.2 **The applied partnering sunflower model (reproduced from Nyström, 2005 with permission for both print and online use from Taylor and Francis Journals (UK))**

based) elements, and goals when discussing the concept. To thoroughly understand what is specific about alliancing, the focus should be on the hard and soft elements, which were identified through a comprehensive literature review. The second contribution was to make use of Wittgenstein's idea of family-resemblance when defining the construction alliancing concept. It was concluded, based on the comprehensive literature review and content analysis, that there are two essential hard (contractual) elements – formal contract and real gain-share/pain-share arrangement – and three necessary soft (relationship-based) elements – trust, long-term commitment, and cooperation and communication in construction alliancing – and that a number of various elements can be added to constitute a specific variant of alliancing. Figure 3.3 shows the alliancing sunflower model containing all the key elements of alliancing while Figure 3.4 indicates the applied alliancing sunflower model with two case studies. It is indicated that this model provides an innovative method to define the vague and versatile concept of construction alliancing in a flexible and structured way. By doing so, industrial practitioners may find the alliancing sunflower model useful in the procurement phase of a building and construction project as a description of the concept and as a common starting point for discussions between a client and a contractor on how to procure a specific alliancing project, thereby avoiding any misinterpretations of what an alliancing project is.

Definitions of relational contracting

There are numerous definitions of RC and this reflects that it is very difficult to give concise and comprehensive explanations of RC. The concept was first defined by Macaulay (1963) as the working relationship amongst the parties who do not often follow the legal mechanism offered by the written contracts, and the parties themselves govern the transactions within mutually acceptable social guidelines. The foundation of RC is often viewed to be based on recognition of mutual benefits and win-win scenarios through more cooperative relationship between parties (Macniel, 1978; Alsagoff and McDermott, 1994; Jones, 2000; Rowlinson and Cheung, 2004b; Kumaraswamy et al., 2005). RC embraces and underpins different approaches, including partnering, alliancing, joint venture and other collaborative working arrangements and better risk-sharing mechanisms (Macneil, 1978; Alsagoff and McDermott, 1994; Jones, 2000; Rahman and

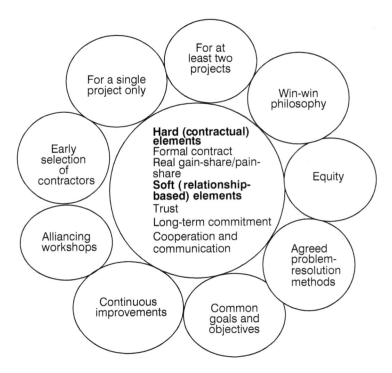

Figure 3.3 Alliancing sunflower model containing all the key elements of alliancing (reproduced from Yeung et al, 2007a with permission for both print and online use from Taylor and Francis Journals (UK))

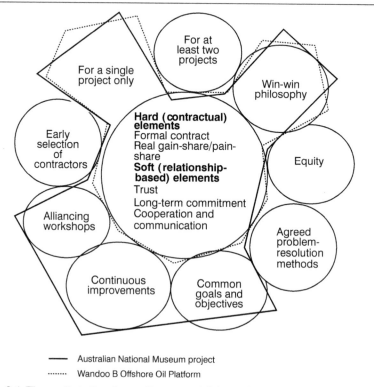

Australian National Museum project
.... Wandoo B Offshore Oil Platform

Figure 3.4 The applied alliancing sunflower model (reproduced from Yeung et al., 2007a with permission for both print and online use from Taylor & Francis Journals (UK))

Kumaraswamy, 2002, 2004; Rowlinson and Cheung, 2004b). Kumaraswamy et al. (2005) advocated that the core of RC is to establish the working relationships between the parties through a mutually developed, formal strategy of commitment and communication aimed at win-win situations for all parties. Sanders and Moore (1992), however, considered that the aim of RC is to generate an organizational environment of trust, open communication and employee involvement.

Palaneeswaran et al. (2003) viewed RC as win-win approaches, which encompass partnering and alliancing to provide vehicles for clients and contractors to strive towards excellence by achieving quality with greater value. Walker and Chau (1999) stated that RC offers a cost-effective means of encouraging collectively beneficial behaviour when transactions are exposed to opportunism, but a fully contingent contract is too costly to specify. The Australian Constructors Association (1999) defined RC as a process to build up and manage the relationships between parties to remove barriers, encourage maximum contribution and allow all parties to achieve success. McLennan (2000) described RC as a way to maximize project outcomes for all parties in the new complicated environment by adopting a conscious approach to build up and manage relationships alongside the cooperative application of ever-improving project delivery systems and processes. Obviously, various researchers put different emphases on the definition of RC and it should not be difficult to observe that some common threads exist although it is too early to be able to derive a single, concise and comprehensive definition of RC. Table 3.1 attempts to summarize the major definitions proposed by these researchers.

Table 3.1 Definitions of RC

Author	Definitions of RC
Macaulay (1963)	The working relationship amongst parties who do not often follow the legal mechanism offered by the written contracts, and the parties themselves govern the transactions within mutually acceptable social guidelines.
Macneil (1978); Alsagoff and McDermott (1994); Jones (2000); Rahman and Kumaraswamy (2002, 2004); Rowlinson and Cheung (2004b); Kumaraswamy et al. (2005)	Based on recognition of mutual benefits and win-win scenarios through a more cooperative relationship between parties. RC embraces and underpins different approaches, encompassing partnering, alliancing, joint venture and other collaborative working arrangements and better risk sharing mechanisms.
Sanders and Moore (1992)	To generate an organizational environment of trust, open communication and employee involvement.
Walker and Chau (1999)	To offer a cost-effective means of encouraging collectively beneficial behaviour when transactions are exposed to opportunism, but a fully contingent contract is too costly to specify.
Australian Constructors Association (1999)	A process to build up and manage the relationships between the parties that aims to remove barriers, encourages maximum contribution and allows all parties to achieve success.
McLennan (2000)	RC is a way to maximize project outcomes for all in the new complicated environment by adopting a conscious approach to build up and manage relationships alongside the cooperative application of ever-improving project delivery systems and processes.
Palaneeswaran et al. (2003)	Win-win' relational contracting approaches, such as partnering and alliancing, provide vehicles for clients and contractors to drive towards excellence by achieving quality with greater value.
Kumaraswamy et al. (2005)	To establish the working relationships between the parties through a mutually developed, formal strategy of commitment and communication aimed at win-win situations for all parties.

Using Wittgenstein's family-resemblance concept to define relational contracting

The numerous definitions of RC mentioned above show how hard it is to give a concise and comprehensive explanation of the concept. It is clear that there is no agreement on which specific elements should be included in defining RC and thus the concepts appear vague and difficult to be compared. In fact, Wittgenstein argued that complex concepts are unable to be defined in the traditional way by stating necessary and sufficient conditions because there may not be a single set of characteristics that are common for all variants of a concept (Nyström, 2005; Yeung et al., 2007a). Instead he regarded that complicated networks overlap similarities amongst the things that fall under a complex concept (Kenny, 1975). Murphy (1991) stated that Wittgenstein's classical example is the term 'game', which is illustrated in such a way that a large number of activities are characterized as games. Nevertheless, he argued that there is not a single and common characteristic for all of them. He further elaborated that ball games such as tennis, football and basketball have rules to be followed. But there are no rules stated clearly when a person just throws a ball in the air. Some elements of the ball games, encompassing rules, competitiveness, and reward and penalty, remain but some fall off, including hard physical work and the ball, when the thought goes to board games. Wittgenstein suggests that there is a complex network of overlapping characteristics without any common features covering all kinds of games. Such an approach to understanding a versatile concept is called 'family-resemblance' because it resembles the type of similarity that is found within a family. He further used the following example to illustrate the family-resemblance concept. A daughter in a family could have the 'same' nose as her father, while the father and his son have the 'same' ears, but there is no feature common to all members of the family. Nevertheless, there is still a bond between them (Kenny, 1975). Clearly, the family-resemblance method defining a multifaceted concept is more flexible and yet structured because it does not restrict the meaning of a concept to a small number of simple characteristics. Therefore, it is suitable to use this innovative method to understand complex concepts, such as RC, that are vague and multifaceted in nature. It should be noted that there are four major forms of RC which are going to be discussed: (1) partnering; (2) alliancing; (3) joint venture (JV); and (4) PPP.

Essential elements of relational contracting

By using the content analysis[2], twelve elements of RC were identified from the analysed literature as shown in Figure 3.5. Based on the reviewed literature, 'commitment', 'trust', 'cooperation and communication', 'common goals and objectives' and 'win-win philosophy' are the most important element in

RC because they are cited with the highest frequencies by the authors. The following sections briefly present the elements that constitute the whole 'RC family' as they are described in the literature. After that, the application of family-resemblance approach to the RC concept is illustrated.

Commitment

A number of researchers stated that, as practiced in the construction industry in its myriad forms, the core of RC is to establish a working relationship amongst different parties through a mutually developed, formal strategy of commitment and communication targeted at win-win situations for all parties (Rahman and Kumaraswamy, 2002a, 2004; Kumaraswamy et al., 2005). Tomlinson (1970) defined a joint venture as an arrangement where there is commitment of funds, facilities and services by two or more legally separated interests, to an enterprise for their mutual benefit for more than a short period of time. Asian Development Bank (2007) opined that the application of PPP and its successful implementation is based on the top-level commitment of the central government and local governments. Senior leaders and municipal mayors have to support the concept of PPP and

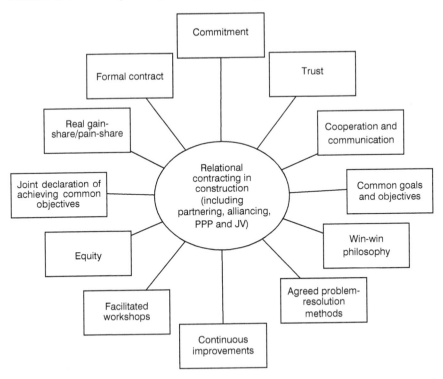

Figure 3.5 Key elements of RC in construction (including partnering, alliancing, PPP and JV)

take a leadership role in PPP projects. In fact, a well-informed leader can help to minimize misconceptions about the principles and merits of PPP. Manley and Hampson (2000) stated that partnering is typically defined in the literature as a commitment between a client and a contractor(s) to actively cooperate to meet separate, but complementary, objectives. Li et al. (2000) viewed that strategic partnering requires long-term commitment and trust by the parties involved to extend their relationships beyond the successful completion of a single project to the formation of an alliance. McGeorge and Palmer (2002) pointed out that in addition to a need for high-level commitment; there is a need for 'internal partnering', while Peters et al. (2001) suggested partnering relies solely on the commitment of individuals due to the fact that the partnering charter is not legally binding. The Construction Industry Institute Australia (1996) stressed that there is no partnering contract and an agreed partnering charter forms the basis of a working agreement that is intended to shape a non-adversarial culture to promote win-win working relationships between partners. This is achieved through fostering cooperative and mutually beneficial relationships amongst project stakeholders and developing an explicit strategy of commitment and communication. These goals are documented in a charter that stands alongside legally binding contractual arrangements (Rowlinson and Cheung, 2004b).

Bennett and Jayes (1998) opined that the strategies need to be supported by all organizations involved in the partnering arrangement, and this commitment gradually builds up through the experience of working together successfully. Hampson and Kwok (1997) proposed that commitment is one of the key characteristics of successful strategic alliances as well as successful business relationships. Walker et al. (2000b) emphasized that trust and commitment underpins the three essential elements of partnering and alliancing (mutual objectives, problem resolution, and continuous improvement). They added that commitment is the physical and mental manifestation of the concept of trust. It means that another party will take this trust on board and 'live up to' the spirit of the bargain by committing more personal pride and obligation to 'do the right thing' than would otherwise be the case. Walker et al. (2002) added that the delivery management plan of the National Museum of Australia project was established on the basis of an alliance concept. The core principle of alliancing was to achieve a positive outcome for all alliance members including the client (also an alliance member) through shared commitment to a common goal of project realization delivering best value to the client and acceptable reward outcomes to alliance members. The assumption made is that all parties can achieve a win-win situation provided that they work together to help each other to gain not only a realistic reward for their input but to gain a competitive edge in the market as a result of their experience on this milestone project. Thorpe and Dugdale (2004) also agreed that successful alliance contracting requires commitment by both parties to achieve common goals. Alchimie and Phillips (2003) found that project alliances are characterized by uncompromising commitments to trust, collaboration, innovation and mutual support to achieve outstanding results. Lendrum (2000) regarded that for the purpose of making alliances successful, all parties have to agree on the objectives and share the principle's process and general information to gain a partner's initial and ongoing support and commitment.

Trust

Many researchers (Construction Industry Institute Australia, 1996; Green, 1999; Li et al., 2000; Lazar, 2000; Sanders and Moore, 1992; Manley and Hampson, 2000; Kumaraswamy et al., 2005; Rahman et al., 2007) viewed trust as a core element of RC. When this element is developed, other sub-elements are likely to be achieved and the benefits to all parties are more easily maximized (Construction Industry Institute, 1991; Bennett and Jayes, 1998). Sanders and Moore (1992) viewed that the aim of partnering is to generate an organizational environment of trust, open communication and employee involvement. Crowley and Karim (1995) mentioned that partnering is typically defined in one of the two ways: (1) by its attributes, such as trust, shared vision and long-term commitment; and (2) by the process where partnering is seen as a verb, as opposed to a noun, and refers to such as developing a mission statement, agreeing on goals, and organizing/conducting partnering workshops. Walker et al. (2000b, 2002) stated that partnering and alliancing are based upon a need for mutual trust to generate commitment and constructive dialogue, and trust is part of an outcome from negotiation. In fact, trust is bound up with past experience, both directly with the persons concerned, and indirectly through projected or anticipated experiences; thus trust is an intensely emotional and human phenomenon. Walker et al. (2002) pointed out that the formation of a partnering and alliance team in a true joint management group with democratic membership ensures that trust and commitment are encouraged and manipulation discouraged by the system of alliancing. Hampson and Kwok (1997) proposed trust as an important element of successful strategic alliances as well as successful business relationships. Howarth et al. (1995) believed that 'no successful strategic alliances can be developed without trust. Trust in a strategic alliance also includes the concept of reciprocity, which implies a long-term focus, the acceptance that obligations are mutual, and room for adjustment if one partner is suddenly placed in a compromising position. Hauck et al. (2004) agreed that trust and integrity are essential for true collaboration. Alchimie and Phillips (2003) viewed project alliancing as an integrated high-performance team selected on a best person for the job basis; sharing all project risks with incentives to achieve game-breaking performance in pre-aligned project objectives; within a framework of no fault, no blame and no dispute; characterized by uncompromising commitments to trust, collaboration, innovation and mutual support – all in order to achieve outstanding results. Kumaraswamy et al. (2005) and Rahman et al. (2007) perceived that trust should be at the core of RC approaches to construction procurement. Rahman and Kumaraswamy (2005) conducted a study and confirmed that trust and business ethics and strategies are more helpful for collaborative working arrangements than some others. Figure 3.6 shows how trust is at the centre of all aspects of a working partnership.

Cooperation and communication

A number of researchers stated that RC is based on recognition of mutual benefits and win-win philosophy through more efficient cooperation and communication between the parties (Macneil, 1978; Alsagoff and McDermott, 1994; Jones, 2000; Rahman and Kumaraswamy, 2002, 2004; Rowlinson and Cheung, 2004b; Kumaraswamy et al., 2005). Rahman and Kumaraswamy (2004) stated that win-win working relationships between partners is achieved by fostering cooperative and mutually beneficial relationships amongst project stakeholders and developing an explicit strategy of commitment and communication. Walker et al. (2000b) stated that partnering and alliancing are founded upon team spirit and the honesty associated with notions of trust, commitment and the application of power and influence. Excellent and effective communication is essential for successful

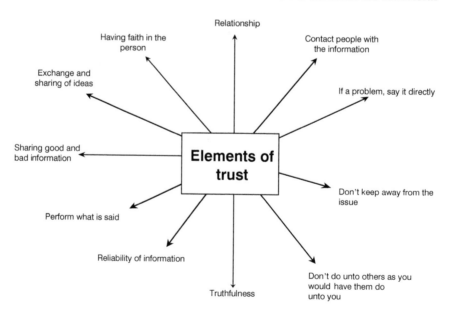

Figure 3.6 Elements of trust (adapted from Whiteley et al., 1998)

relationship building. Construction Industry Institute (1991) also viewed effective communication and cooperation as a vital partnering element. Hampson and Kwok (1997) stressed that cooperation and communication is a key element of successful alliances. Abrahams and Cullen (1998) opined that cooperative working between entities is an important element for alliancing parties to succeed. Both Hauck et al. (2004) and Walker et al. (2002) pointed out that the intense integration of alliance partners through the whole collaborative process requires excellence in communication at a personal level, at a business level and at an operational level. This generally requires a quantum leap in the use of shared information technology (IT) systems and information processing integration. Alchimie and Phillips (2003) also agreed that cooperation and collaboration are vital elements for successful alliances. Walker and Johannes (2003) observed that large-scale construction infrastructure projects in Hong Kong are always delivered using joint venture of global and local construction organizations. They found that reasons for companies to do so include bridging knowledge and expertise gaps, sharing risk, and exploring opportunities to add value to the JV organizations through collaboration and cooperation.

Common goals and objectives

Walker and Chau (1999) reckoned that RC approaches appear useful in achieving common goals and objectives, which are to reduce production and transaction costs. As mentioned previously, recognition of mutual benefits and win-win scenarios are the two bases of RC and they are achieved by more cooperative relationships between parties. This implies that they often have common agenda to achieve them. Construction Industry Institute (1991) viewed shared vision (common goals and objectives) as a vital partnering element in which each of the partnering organizations must understand the need for a shared vision and common mission for the partnering relationship. Bennett and Jayes (1998) proposed that one of the three key elements of partnering is to agree common goals to take into account the interests of all the firms involved. Crowly and Karim (1995) agreed that a typical partnering definition is based on its attributes, including (1) trust; (2) shared vision (common goals and objectives); and (3) long-term commitment. Partnering is about people within partnered organizations making a commitment and building trust to work together towards their common project goals and objectives (Walker et al., 2000b, 2002). In a PPP project, it is often mutually agreed to deliver the project on-time, with enhanced asset quality and cost reductions through the equitable allocation of risks (Asian Development Bank, 2007). Shaughnessy (1995) viewed that the most vital prerequisite for success in an international joint venture is that the parties should share the same objectives. Rowlinson and Cheung (2004b) pointed out that partnering is defined as a structured management approach to facilitate team working across contractual boundaries. Its fundamental elements include (1) mutual objectives; (2) agreed problem-resolution methods; and (3) an active search for continuous measurable improvements. Project alliancing is described as a cooperative arrangement between two or more organizations that forms part of their overall strategy and contributes to achieving their common goals and objectives for a specific project (Kwok and Hampson, 1996). Walker et al. (2000b, 2002) explained that the core principle of alliancing was to achieve a positive outcome for all alliance members through shared commitment to common goals and objectives of a project realization to deliver best value to the client and acceptable reward outcomes to alliance members. Thorpe and Dugdale (2004) viewed that alliance contracts are best suited to contracts that require innovation and commitment to achieve common goals. Hauck et al. (2004) also agreed that common goals and objectives are key elements for successful alliance contracts.

Win-win philosophy

Similar to common goals and objectives, win-win philosophy is a key element for RC as recognition of mutual benefits and win-win scenarios are often viewed as a basis of the foundation of RC and they are achieved through more a cooperative relationship between parties (Macniel, 1978; Alsagoff and McDermott, 1994; Jones, 2000; Rowlinson and Cheung, 2004b; Kumaraswamy et al., 2005). Crowley and Karim (1995) viewed win-win philosophy as an important element for partnering, and they defined it as 'neither party wins due to the other's losses'. Lazar (2000) mentioned that partnering is able to guide people both on- and off-site into the types of interactions and relationships that produce a win-win outcome. Lendrum (2000) mentioned that in order for alliances to be successful, all parties have to agree on the objectives and share the principles, processes and general information to gain their partner's initial and ongoing support and commitment. The contractor must be involved to ensure a win-win long-term relationship.

Geringer (1988) stated that joint ventures involve two or more legally distinct organizations, and each of them shares in the decision-making activities of the jointly owned entity to achieve a win-win situation. Walker and Johannes (2003) observed that large-scale construction infrastructure projects in Hong Kong have a tendency to use joint venture to achieve win-win outcomes, including bridging knowledge and expertise gaps, sharing risk and exploring opportunities to add value to the JV organizations. Walker et al. (2002) stated that an element of alliances was that joint budget and cost-and time-committed targets, established through an alliance board are represented by key senior project champions from each alliance member and the owner/client. Abrahams and Cullen (1998) defined alliancing as 'an agreement between entities which undertake to work cooperatively, on the basis of a sharing of project risk and reward, for the purpose of achieving agreed outcomes...', implying a win-win philosophy. Asian Development Bank (2007) perceived that PPP can achieve win-win results for all parties concerned because a core principle of PPP arrangement is the allocation of risk to the party best able to manage or control it. In general, the private sector is more efficient in asset procurement and service delivery because of its market-driven orientation. It is, therefore, to the government's advantage to share the associated risks with the private-sector participants.

Agreed problem-resolution methods

Agreed problem-resolution methods are considered as an important elements for RC, especially for both partnering and alliancing. As suggested by Bennett and Jayes (1998), one of the three key elements of partnering is to make decisions openly and to resolve problems in a way that was jointly agreed at the start of a project. Walker et al. (2002) stated that agreed problem resolution is essential when establishing trust and commitment between parties. The Construction Industry Institute task force considered that a successful partnering relationship element included conflict resolution through agreed problem solving (Crowley and Karim, 1995). Rowlinson and Cheung (2004b) agreed that a fundamental element for partnering is agreed problem-resolution methods. Walker et al. (2000b) stressed that the three essential elements of partnering and alliancing – (1) mutual objectives; (2) agreed problem resolution; and (3) continuous improvement – are underpinned by trust and commitment. Problem- and dispute-resolution procedures adopted in alliancing provide for the types of problems to be defined and reasonable timeframes for resolution stipulated. The reason for escalating a dispute may be hardening of diverse positions or may simply be a result of the party not being authorized to commit required resources to resolve the dispute. In cases where a dispute is escalated unnecessarily, the person escalating the dispute may not be appreciated by his peer groups. This provides a self-regulating mechanism for ensuring that problems are indeed resolved at the lowest possible level. Hampson and Kwok (1997) also proposed that joint problem-solving method is a key element of the successful alliances. A standard problem resolution flow chart is shown in Figure 3.7.

Continuous improvements

A key element for RC, particularly partnering and alliancing (Construction Industry Institute, 1991; Bennett and Jayes, 1998; Walker et al., 2000b) is continuous improvement, meaning that long-term targets are set and achieved by all the stakeholders. Rowlinson and Cheung (2004b) agreed that a fundamental element for successful partnering encompassed an active

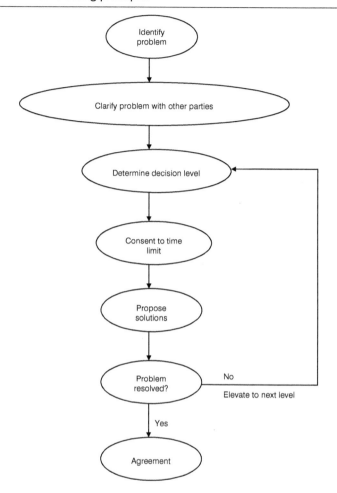

Figure 3.7 Problem resolution flow chart (reproduced from Bennett and Jayes, 1995 with permission for both print and online use from the School of Construction Management and Engineering, The University of Reading)

search for continuous measurable improvements. Cheng and Li (2004a) also agreed that continuous improvement is a vital element for successful strategic partnering to create a good learning culture. Garvin (1993) illustrated that continuous improvement involves continuous learning devoted to gradual process improvement (total quality management, TQM), radical process improvement (business process reengineering, BPR) and learning process improvement (a learning organization) (Kilmann, 1995). Walker et al. (2000b, 2002) observed that an essential element of partnering and alliancing was continuous improvement; performance is measured and analysed to provide knowledge about how improvement can be achieved continuously. There must be a commitment to learn from experience and to apply this knowledge to improve performance. Thorpe and Dugdale (2004) found that a vital element of alliance was continuous improvement to achieve results on time and to full specification requirements, while stating that innovation will always be required to improve the current process. Under private finance initiative (PFI), a major type of PPP, the public sector only states the requirements of the service outcomes but not prescriptive facilities. This allows the private sector to innovate in the ways it provides for those outcomes. The management skills and innovations of the private sector can lead to reduced project costs, increased efficiencies and continuous improvements (The International Project Finance Association, 2001).

Facilitated workshops

Facilitated workshops are also key elements for partnering and alliancing but with relatively less importance. Green (1999) considered that partnering workshops need to be continuous and not a one-off at the project start. Walker et al. (2000b, 2002) pointed out that the interviewing process to derive a shortlist of potential alliance members requires sophistication and judgment of a client as do the facilitated workshops. This means that alliancing workshops are a useful tool to help select capable construction alliance partners.

Equity

Equity is another foundation for successful partnering and alliancing implementation. All the interests of stakeholders should be considered in creating mutual goals and there should be commitment to satisfying each stakeholder's requirement based on equity (Crowley and Karim, 1995; Li et al., 2000; Lazar, 2000; Walker et al., 2002). It reflects a sense of proportionality and balance transcending simple fairness (Construction Industry Institute Australia, 1996). Bennett and Jayes (1998) opined equity as one of the seven pillars for the second generation partnering, and it can be defined to ensure each contracting party is rewarded for his work based on fair prices and fair profits. Manley and Hampson (2000) showed that one of the alliancing features is an equitable risk-reward balance that aligns the commercial interests of the parties. Hauck et al. (2004) agreed that the foundation of the collaborative process for project alliancing is equity between parties.

Joint declaration to achieve common objectives

Many researchers stated that there are generally some important statements highlighting a joint declaration between parties to achieve shared objectives in different forms of RC. For instance, there are always some key statements mentioning common objectives in a partnering charter built up by different project stakeholders for adopting structured partnering approach (Construction Industry Institute 1991; Construction Industry Institute Australia 1996; Manley and Hampson 2000). For alliancing, there are also some joint declaration statements highlighting some common objectives in an alliance contract (Walker et al., 2000b, 2002). For PPP, it often generally includes statements containing common objectives to achieve value for money and reduce project risks by transferring them to parties best able to manage them (Asian Development Bank, 2007; Li et al., 2005). For JV, there is usually a joint declaration to achieve common objectives, i.e. technology transfer (Carrillo, 1996).

Real gain-share/pain-share

Walker et al. (2000b, 2002) analysed that with alliancing, there is a 'joint' rather than 'shared' commitment. Parties agree their contribution and required profit levels beforehand and then place these levels at risk. If one party in the alliance under-performs, then all other alliance partners are at risk of losing their rewards (profit and incentives) and could even share losses according to the agreed project gain-sharing/pain-sharing model. Abrahams and Cullen (1998) defined project alliances as an agreement between entities which undertake to work cooperatively, based on a mechanism of project risk and reward sharing to achieve agreed outcomes. This approach is based on principles of good faith and trust as well as an open-book accounting approach towards costs. Hauck et al. (2004) mentioned that as an alliance of talented professionals pooling resources to achieve the project goal, they develop the project price target through design development with agreed risk and reward sharing arrangements like guaranteed maximum price (GMP) and target cost contracting (TCC) procurement strategies. The parties agree on a risk and reward formula where an open-book accounting approach is undertaken to determine cost reimbursement together with agreed and verified site management costs to establish a base target cost. The firm's corporate profit (usually determined from audited figures over an agreed period) is placed as an 'at risk' element to ensure that the agreed project costs are met. A bonus reward mechanism to be shared by all parties is jointly established to encourage further innovation and excellence. Therefore, the agreed project cost can only be determined when the alliance partners have been selected. McGeorge and Palmer (2002) emphasized that alliancing differs radically from partnering with regard to risk and reward sharing. In partnering the client still ultimately purchases a product (usually a building) which is produced, albeit in a spirit of mutual cooperation, with the design and construction team. In alliancing, the virtual corporation produces the product with each member of the corporation sharing risks and rewards. The characteristics of successful strategic alliances proposed by Hampson and Kwok (1997) are trust, commitment, interdependence, cooperation, communication and joint problem solving. The interdependence here implies sharing risks and rewards. A classical alliancing gain-share/pain-share scheme is shown in Figure 3.8.

Formal contract

McGeorge and Palmer (2002) viewed that alliancing is somewhat akin to the slogan of the three musketeers 'All for one and one for all' in that alliancing could be described as partnering underpinned with economic rationalism given that alliance partners coalesce into a virtual corporation in which agreed profit and loss outcomes are contractually binding on all parties (Walker et al., 2000a, 2000b). Rowlinson and Cheung (2004b) pointed out that a project alliancing agreement is legally enforceable while Hauck et al. (2004) also stated that the project alliancing 'agreement' is a legally binding contract and, therefore, legally enforceable.

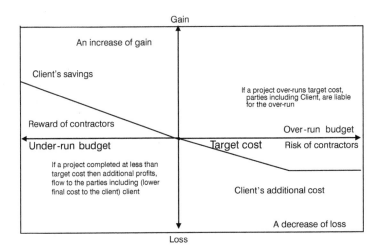

Figure 3.8 Typical model of gain-sharing/pain-sharing philosophy (reproduced from Australian Constructors Association, 1999 with permission for both print and online use from Australian Constructors Association

Analysing relational contracting using Wittgenstein's family-resemblance concept

Twelve common elements were identified from the literature in the study of RC (see Figure 3.5). 'Commitment', 'trust', 'cooperation and communication', 'common goals and objectives', and 'win-win philosophy' appear to be the most vital elements for RC because they have been cited by most researchers when defining the RC concepts, and are therefore interpreted as core elements for RC. The family-resemblance philosophy proposed by Wittgenstein was adopted to define RC in construction by first identifying core elements for RC in construction (acting together as the centre of a sunflower) and a number of non-core elements (acting as petals of the sunflower). Different combinations of core elements and non-core elements constitute different variants of RC in construction.

The resulting analysis of the RC concept can be described as a 'sunflower' because there must be a centre containing the five common core elements to all RC designs, combined with the other non-core elements as mentioned in Figure 3.5. These elements can be seen as petals of the sunflower. A contracting practice can be defined as a type of RC if it first contains the five core elements and second, some of the petals. It should be noted that there is no specific petal or a set of petals that the RC sunflower has to contain. Therefore, adding different sets leads to variants of RC. The sunflower as an entity can be seen as the base for portraying the whole 'family' of all RC variants (Figure 3.9).

Application of Wittgenstein's family-resemblance concept to relational contracting

The above-mentioned structure facilitates a practical application of the somewhat vague and multifaceted concept of family-resemblance. Various designs of RC projects can be captured within the same structure, which is indicated by the following four case studies.

Case 1: Chater House (a partnering project in Hong Kong)

The first case is taken from Chan et al. (2004a, b, c; 2006), where they described the Chater House, a prestigious office development project in Hong Kong. The project was composed of a demolition of an existing building (Swire House) and a construction of a 29-storey international Grade A office development in Hong Kong's central business district, which comprised of a 3-storey basement, a 3-storey podium and a 23-storey tower. The overall gross floor area was 74,000m². The contract sum was approximately HK$1.2 billion (approximately US$154 million), with original contract duration of 641 days. The project was procured by a negotiated guaranteed maximum

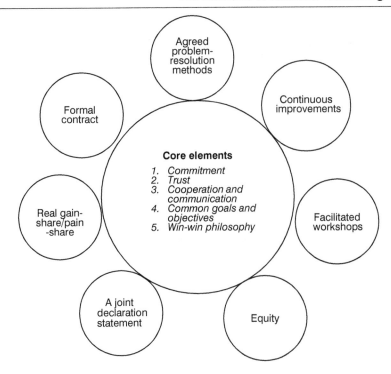

Figure 3.9 RC sunflower model containing all the key elements of RC (adapted from Nyström, 2005 and Yeung et al., 2007a)

price (GMP) contract together with adopting a partnering approach. The mechanism of the GMP contract requires that the client, consultants and the main contractor work as a team to determine construction method, programme, pricing, detailed breakdown of direct works, and consent to preliminaries and conditions of contract. This entailed the main contractor to release all his back-up data in an open manner to team members. The exchange of this information required a high level of trust amongst the team, especially the main contractor. It was reported (Chan et al., 2004a) that support from senior management and commitment from all participants were two of the most critical success factors for achieving partnering success in this project. In addition, the interviewees perceived that the level of trust was high and there was a good collaborative working environment in this project. The reasons were partly because there were clear common goals and objectives mentioned in the partnering charter and parties tried hard to achieve a win-win philosophy between them.

In addition to these five core elements, namely, 'commitment', 'trust', 'cooperation and communication', 'common goals and objective' and 'win-win philosophy', this RC project also included: (1) partnering charter; (2)

agreed problem-resolution methods; and (3) facilitated workshops. The variant of RC is shown by the set of elements within the thin solid line boundary in Figure 3.10.

Case 2: Australian National Museum project (an alliancing project in Australia)

The second case is found from Walker et al. (2000b, 2002) and Hauck et al. (2004) where the Australian National Museum project was described. This was a large-scale public sector project with the contract sum of approximately AU$150 million. The project began in 1998 and was completed in 2000. There were about 300 participants involved in this project, in which there were 10 from the client, 6 from the main contractor, 15 from the consultants, and 260 from the subcontractors. The client was the National Museum, the

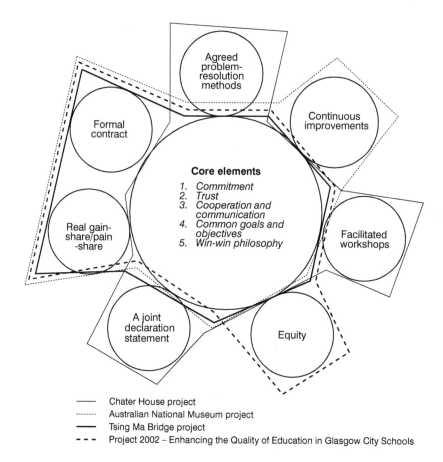

——— Chater House project

········· Australian National Museum project

▬▬▬ Tsing Ma Bridge project

– – – Project 2002 – Enhancing the Quality of Education in Glasgow City Schools

Figure 3.10 The applied RC sunflower model

main contractor was Bovis Lend Lease, and a main consultant was Peck Von Hartel (architect).

It was found from Hauck et al. (2004) that support from senior management and commitment to alliancing from project participants were two of the most critical success factors for achieving alliance success in this project. A collaborative process included integrity and trust which were essential for true collaboration in this alliance project. In fact, the trust level was high and there were good collaborative teams in this project. The communication between parties was good and they worked jointly toward a common goal. The Collaborative Process Institute (CPI) concluded that collaborative communication had to be based on key principles, which encompass equality, openness, problem orientation, positive intent, empathy and extensive use of technology. Apart from the five core elements, this RC project also included: (1) formal contract; (2) real gain-share/pain-share; (3) continuous improvements; and (4) facilitated workshops. The variant of RC is delineated by the set of elements within the dotted line boundary in Figure 3.10.

Case 3: Project 2002 – Enhancing the quality of education in Glasgow city schools (a PPP project in the UK)

The third case is found from the website of Hong Kong Efficiency Unit (2008) (www.eu.gov.hk/english/case/case.html) where Project 2002 – Enhancing the quality of education in Glasgow city schools was mentioned. There were a total of 39 secondary schools in Glasgow, Scotland in 1996. A majority of them were in poor condition, and they needed substantial refurbishment with accumulative maintenance costs estimated to be more than £100m. In spite of the fact that the schools could accommodate more than 50,000 pupils, there were only about 29,000 pupils in the city. It was expected that this figure would rise only slightly in the near future. The sponsor for the project was Glasgow City Council, Scotland. The consortium was 3Ed (Miller Group Limited, Amey Ventures Ltd and Halifax Projects Investments Ltd). The capital cost was £225m, with the contract length of 30 years and information communications technology (ICT) of 12 years.

Commitment from the consortium was a critical success factor for achieving construction success in this project. In addition, it was perceived that the level of trust was quite high and project team members worked collaboratively and harmoniously in this project. It was partly because they had clear common goals and objectives, i.e. achieving value for money, innovative financing, risk transfer, and change management. Besides, different parties endeavoured to achieve win-win outcomes for themselves. In addition to the five core elements, this RC project also included: (1) formal contract; (2) real gain-share/pain-share; and (3) equity. The variant of RC is delineated by the set of elements within the dash line boundary in Figure 3.10.

Case 4: Tsing Ma Bridge (a joint venture project in Hong Kong)

The fourth case is found from the websites of Wikipedia (http://en.wikipedia. org/wiki/Tsing_Ma_Bridge) and the Hong Kong Polytechnic University (www.cse.polyu.edu.hk/~ctbridge/case/tsingma.htm). The Tsing Ma Bridge is a suspension bridge in Hong Kong, China and is named after two of the islands it connects, *Tsing* Yi and *Ma* Wan. The Tsing Ma Bridge is the seventh longest-span suspension bridge in the world. The client was the Highways Department of the Hong Kong SAR Government; the main contractor was Anglo Japanese Construction JV; and the engineer and designer was Mott MacDonald Hong Kong Ltd. It has two decks and carries both road and rail traffic. The main span of the bridge is 1,377 metres (4,518 ft) long and 206 metres (676 ft) high. The span is the largest of all bridges in the world that carry rail traffic. The 41 metres (135 ft) wide bridge deck carries six lanes of automobile traffic, three lanes in each direction. There are two rail tracks in the lower level. In addition, there are two sheltered carriageways on the lower deck for both maintenance access and as backup for traffic in case typhoons strike Hong Kong. Although road traffic would need to be closed in that case, trains could still get through in either direction. Table 3.2 shows the details of this JV project.

It is found that apart from the five core elements, this RC project also included: (1) formal contract; and (2) real gain-share/pain-share. The variant of RC is delineated by the set of elements within the thick solid line boundary in Figure 3.10.

Significance and value of relational contracting sunflower model

An increasing number of client organizations have been observed to adopt RC to manage their building and construction works over the past decade (Walker et al., 2000a; Chan et al, 2002a). With time, the development of RC becomes complicated, and it is quite difficult to define what a construction RC project is. In fact, industrial practitioners and academics are always vague with the concepts and definitions of RC. By the adoption of Wittgenstein's idea of family-resemblance, an RC sunflower model has been proposed. The model provides an innovative and useful framework to define the vague and versatile concept of RC in construction in a more flexible, yet structured, way. Industrial practitioners may find the RC sunflower model useful in the procurement phase of a construction project, particularly if needed, both as a description of the concept and as a common starting point for discussions between the client and the contractor on how to procure a RC project. With a better understanding of this complicated concept, it could help to identify critical success factors for RC projects and develop a best practice framework for managing future RC projects to strive for construction excellence.

Table 3.2 Details of the Tsing Ma Bridge project
(adopted from http://en.wikipedia.org/wiki/Tsing-Ma-Bridge)

Official name	Tsing Ma Bridge
Client	Highways Department of the Hong Kong SAR Government
Engineer and designer	Mott MacDonald Hong Kong Limited
Main contractor	Anglo Japanese Construction JV
Construction date	May 1992
Completion date	May 1997
Opening date	27 April 1997
Construction cost	HK$7.2 billion
Toll	HK$30 (cars)
Carries	6 lanes of roadway (upper); 2 MTR rail tracks; 2 lanes of roadway (lower)
Crosses	Ma Wan Channel
Locale	Ma Wan Island and Tsing Yi Island
Design	Double-decked suspension bridge
Longest span	1,377 metres (4518 ft)
Width	41 metres (135 ft)
Height of towers	206 metres
Vertical clearance	62 metres (203 ft)

Chapter summary

Based on an in-depth analysis of reported literature, this chapter has tried to make a significant contribution to define RC in construction by using the innovative and useful sunflower model. Wittgenstein's ideas are that complicated concepts can be understood as a network of overlapping similarities. This is dissimilar to the traditional definition whereby a concept is given necessary and sufficient conditions. The RC literature was examined according to the Wittgenstein's philosophy and it was found that five core elements were always included in descriptions: 'commitment', 'trust', 'cooperation and communication', 'common goals and objective', and 'win-win philosophy'. Besides these core elements, there was an overlapping network of the other elements. The contribution can be of paramount importance to both the research community and the construction industry. Previous research on adopting the same research methodologies to define construction partnering and alliancing have proved that this method to define RC in construction is valid and sound. In Chapter Four, the benefits of RC will be discussed.

Chapter 4

Benefits of adopting relational contracting

Introduction

It is widely accepted that the construction industry has suffered from little cooperation, limited trust, fragmentation and ineffective communication and cooperation, thus often inducing a confrontational working relationship between different parties (Moore et al., 1992; Chan et al., 2004b). RC has been introduced to be an innovative, non-adversarial, relationship-based approach to the procurement of construction services in many nations, such as the USA, the UK, Australia and Hong Kong over the last decade (Rahman and Kumaraswamy, 2002b; Palaneeswaran et al., 2003; Kumaraswamy et al., 2005). In fact, RC is an approach to manage complex interpersonal relationships between various players in construction projects. Mutual benefits and win-win scenarios are the two foundations of RC and they are achieved by more cooperative relationships between parties (Macneil, 1978; Alsagoff and McDermott, 1994; Jones, 2000; Kumaraswamy et al., 2005; Rowlinson and Cheung, 2004c). Kumaraswamy et al. (2005) opined that RC helps to establish the harmonious working relationships between parties through a mutually developed, formal strategy of commitment and communication, ultimately aiming at win-win situations for all parties concerned.

Research on the benefits of adopting different forms of relational contracting

Relationship management in Queensland Department of Main Roads

Rowlinson and Cheung (2004) conducted a research study on RC in Queensland Department of Main Roads (QDMR). The authors identified a number of major benefits of adopting the RC approach, including: (1) less paperwork; (2) more enjoyable work environment; (3) people are more helpful, less destructive and more proactive; (4) QDMR can potentially make savings in their operations; (5) QDMR becomes more proactive in helping the contractor; and (6) training and education.

An empirical study of the benefits of construction partnering in Hong Kong

Chan et al. (2003a) reviewed the partnering literature in the area of construction and endeavoured to construct a comprehensive picture of benefits for the partnering practice. It is observed that by adopting a partnering approach, all key project parties are more actively involved so that the project is more likely to be completed on schedule, within budget, and with the fewest conflicts, claims and work defects. The authors also reported on the empirical findings of a questionnaire survey of the benefits of adopting partnering in the Hong Kong construction industry. The findings showed that the top ten partnering benefits were: (1) improved relationship amongst project participants; (2) improved communication amongst project participants; (3) more responsive to the short-term emergency, changing project or business needs; (4) reduction in disputes; (5) better productivity was achieved; (6) a win-win attitude was established amongst the project participants; (7) a long-term trust relationship was achieved; (8) reduction in litigation; (9) improved corporate culture amongst project participants; and (10) improved conflict-resolution strategies.

Introduction to project alliancing on engineering and construction projects

Ross (2001) viewed that alliancing can bring benefits to owners and non-owner participants. The main alliancing benefits to the owners include: (1) much greater certainty of on-schedule or early delivery, particularly in the face of adversity; (2) the project to be delivered very close to or under the agreed target cost; (3) more informed decisions on technical solutions/ choice of equipment; (4) better balance between capital investment and whole-life costs; (5) outcomes that meet or exceed expectations in non-cost areas; (6) potential for real breakthroughs in some areas; and (7) much greater job satisfaction/professional development for all parties involved. On the other hand, alliancing is in general attractive to non-owner participants because: (1) potential for very good returns within acceptable limits of risk; (2) reputation enhancement leading to increased prospects of repeated and referred work; (3) strengthening of relationship with owners and the other participants, thus forming the basis for possible future strategic alliances; (4) increased job satisfaction for staff with associated benefits to overall organizational culture; and (5) significant increase in communication and general project management skills.

Application of public-private partnerships in urban rail-based transportation project

Asian Development Bank (2007) conducted a research study on the application of PPP in urban rail-based transportation project. By conducting a comprehensive literature review and using content analysis, eleven key benefits of PPP were identified: (1) achieving substantial risk sharing; (2) cost savings; (3) value for money; (4) innovations in public services; (5) better maintenance of assets; (6) time savings; (7) encouraging cooperation; (8) enhancing social development and business opportunities; (9) cost certainty; (10) reduced public funding; and (11) time certainty.

Carrillo (1996) conducted a research study on technology transfer on JV projects in developing nations. After interviewing eight UK contractors, he summarized the perceived advantages of JV from both developed and developing countries. The former included: (1) risk spreading; (2) fixed-cost sharing; (3) pooling of knowledge and resources; and (4) reduction of cultural problems. The latter encompassed: (1) balance sheet substance; (2) improved track record; (3) risk sharing; (4) opportunity to work on large projects; and (5) improved purchasing power.

Table 4.1 indicates the summary of previous related research works on investigating the major benefits of adopting different forms of RC. The results show that better cost control was the most frequently cited major benefits for various forms of RC, with better working relationship being the second; and sharing of risk, better time control, efficient problem solving, and potential for innovation being the third.

Benefits of adopting relational contracting

RC can benefit clients, main contractors, consultants, sub-contractors and on-site employees (Palaneeswaran et al., 2003). The key elements of different forms of RC (encompassing partnering, alliancing, PPP and JV), which include commitment, trust, cooperation and communication, common goals and objectives, and win-win philosophy, are planned to include appropriate consideration of the interests of all parties at each level (Construction Industry Institute, 1991; Cowan et al., 1992). The RC process empowers all project personnel to accept responsibility and to do their jobs by delegating decision making and problem solving to the lowest possible level of authority (Kumaraswamy et al., 2005).

After conducting a comprehensive literature review on the perceived major benefits of different forms of RC, they can be grouped under eighteen headings, namely, better cost control, sharing of risk, better quality product, potential for innovation, better time control, better working relationships, lower administrative costs, increased satisfaction, reduced litigation, efficient problem solving, enhanced communication, better safety performance,

Table 4.1 Summary of research investigating the major benefits of adopting different forms of RC

Benefits of adopting different forms of RC	RC (Rowlinson and Cheung, 2004)	Partnering (Chan et al., 2003a)	Alliancing (Ross, 2001)	PPP (Asian Development Bank, 2007)	JV (Carillo, 1996)	Total
1. Better cost control	√	√	√	√	√	5
2. Better working relationship	√	√	√	√		4
3. Sharing of risk			√	√	√	3
4. Better time control		√	√	√		3
5. Efficient problem solving	√	√	√			3
6. Potential for innovation			√	√	√	3
7. Better quality product		√		√		2
8. Enhanced communication		√	√			2
9. Increased satisfaction	√		√			2
10. Drivers for the economy				√	√	2
11. Continuous improvement	√					1
12. Reduced litigation/dispute		√				1
13. Better productivity		√				1
14. Win-win attitude among the project participants		√				1
15. Improved culture					√	1
16. Lower administrative cost	√					1
17. Enhancement of reputation			√			1
18. Reduction of public financing				√		1
Total	6	9	9	8	5	

improved culture, continuous improvement, enhancement of reputation, reduction of public financing, and drivers for the economy. Figure 4.1 shows the major benefits of adopting RC.

Better cost control

Walker and Chau (1999) viewed that it is useful for RC approaches to achieve the overall objective, which is to reduce the sum of production and transaction costs. RC offers a cost-effective means of encouraging collectively beneficial behaviour when transactions are exposed to opportunism. Gransberg et al. (1999) reckoned that partnering, as a major form of RC, has a great potential to improve cost performance. Albanese (1994) opined that the improved cost performance can be considered as an interim benefit resulting from partnering. A number of researchers stated that partnering can help to reduce risk of budget over-runs due to improved cost control (Cowan et al., 1992; Moore et al., 1992; Abudayyeh, 1994;

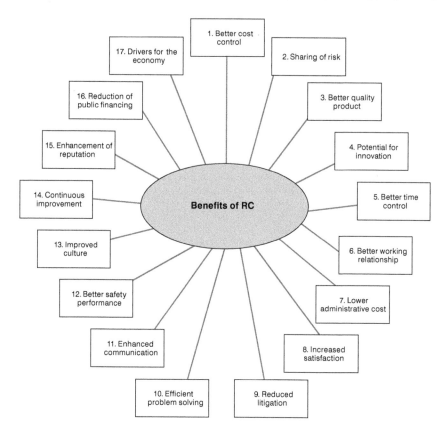

Figure 4.1 Summary of RC benefits

Bates, 1994; Brown, 1994; Back and Sanders, 1996; Construction Industry Institute, 1996; Hellard, 1996b; Ruff et al., 1996; Li et al., 2001; Chan et al., 2003a). Chadwick and Rajagopal (1995) perceived that working with suppliers can enhance the ability of the organization so as to meet the client's programme, quality, flexibility and cost requirements. As a matter of fact, a key benefit of partnering with suppliers is the resultant synergy, thus enabling constant improvement in terms of cost, time and quality. Lewis (1995) identified four key benefits from partnering with suppliers: (1) higher profit margins; (2) lower costs; (3) better value for customers; and (4) a larger market share. Other improvements identified by Lewis included (1) quality improvement; (2) design-cycle time reductions, and (3) increased operating flexibility. Carrillo (1996) stated that JV can assist different parties to share fixed costs for partners from developed nations. Ross (2003) opined that alliancing can assist in delivering the project with optimum outturn cost at or below target outturn cost because of much more effective management of stakeholder issues. And it has great potential for very good returns within acceptable limits of risk. Albanese (1994) stated that the are quite a number of reasons for better cost performance, including: (1) alleviating rework; (2) reducing scheduled time; (3) heightening involvement of team members; (4) improving trust; (5) reducing scope of definitional problems; (6) opening communication; (7) lowering change order rates; (8) improving problem solving; (9) eliminating blame shifting; and (10) improving understanding of project objectives and decreasing adversarial relations. By studying case studies, Asian Development Bank (2007) concluded that procurement methods of the public sector projects under properly structured PPP can be delivered on schedule, with high quality and result in overall cost savings to the project.

Sharing of risk

Gattorna and Walters (1996) argue that the main reason for using a partnering approach to procurement is to develop joint strategies that will achieve strategic objectives. This will enable organizations to improve the return on limited resources while also mitigating their risk. Ng (1997) stated that a major benefit in implementing JV for construction projects is risk sharing between parties. With major projects, a single firm might struggle to raise the financial resources required; through JV, construction firms can combine resources and expertise and therefore increase the likelihood of success. Carrillo (1996) perceived that JV can assist to spread risks for partners from developed countries and share risks for partners from developing nations. Partnering enables parties to share the benefits and resources of other parties and develop management and technical advances together (Cook and Hancher, 1990; Cowan et al., 1992; Li et al., 2001; Chan et al., 2003a). It also establishes the tools for both measurement and sharing of gains and risks

(Ellison and Miller, 1995; Li et al., 2001). Asian Development Bank (2007) stated that the allocation of risk is a core principle of PPP arrangement in that it should be assigned to the party that is best able to manage or control it. In general, the private sector is more efficient in asset procurement and service delivery because of its market-driven orientation. Therefore, it is likely for the government to benefit from sharing the associated risks with private sector participants.

Better quality product

Palaneeswaran et al. (2003) reckoned that win-win RC approaches such as alliancing and partnering enable clients and contractors to achieve construction excellence by achieving high quality with greater value, rather than they could achieve when they try to seize more from each other in a confrontational working atmosphere based purely on the enforcement of binding legal obligations. Partnering produces high quality construction products and services (Construction Industry Institute, 1991; Moore et al., 1992; Bates, 1994; Construction Industry Institute, 1996; Hellard, 1996b; Matthews et al., 1996; Black et al., 2000; Li et al., 2001; Chan et al., 2003a), and engineering rework is reduced in partnered projects (Construction Industry Institute, 1991; Thompson and Sanders, 1998).

An effective partnering agreement improves project quality by replacing the potentially traditional adversarial working relationship and case building with an atmosphere that develops a team building approach to achieve a series of common goals and objectives (Cook and Hancher, 1990; Brown, 1994; Schultzel, 1996; Chan et al., 2003a). Albanese (1994) further explained that the partnering process facilitated communication of quality issues, enabled earlier recognition of potential problems, and helped develop a quality consciousness. Many firms in Albanese's (1994) study were those that practised total quality management, a management approach that usually involved the use of partnering. Asian Development Bank (2007) stated that in a PPP project, the private sector is responsible for ensuring that the asset procured and service delivered meet agreed quality benchmarks throughout the whole life of the contract. The arrangements always encompass incentive payments to reward higher level of performance and quality of service delivered. Therefore, it is often found that the private sector procures the asset with a high quality standard.

Potential for innovation

An effective partnering relationship encourages partners to evaluate advanced technology for its applicability (Cook and Hancher, 1990; Hellard, 1996b; Pena-Mora and Harpoth, 2001). Appropriate use of innovation through open and effective communication improves design and construction processes (Abudayyeh, 1994; Bourn, 2001). Carrillo (1996) reckoned that JV facilitates to pool knowledge and resources for partners from developed countries, thus resulting in higher level of technology transfer and better overall project performance.

Better time control

Bresnen and Marshall (2000c) stated that partnering can lead to dramatically improved project performance in terms of time, cost, fitness-for-purpose, buildability and other criteria. Rowlinson and Cheung (2004c) reckoned that by building a high level of trust and being convinced of contractors' competence and trustworthiness, an organization's personnel can spend more time on creative issues and more focus can be given to creating an excellent project. In fact, many construction projects are completed on schedule as a result of the partnering process (Construction Industry Institute, 1996; Schultzel, 1996). The fair and equitable attitude resolves many disputes, discrepancies and changes conditions which arise during construction. Partnering can reduce the delay as a result of better schedule performance (Cowan et al., 1992; Moore et al., 1992; Ruff et al., 1996; Thompson and Sanders, 1998; Black et al., 2000; Chan et al., 2003a), timely decisions (Albanese, 1994; Li et al., 2001) and reliable programming (Hellard, 1996b; Matthews et al., 1996; Weingardt, 1996). Gransberg et al. (1999) also found that fewer numbers of liquidated damages were imposed on the partnered projects than the non-partnered ones. Ross (2003) perceived that alliancing can help to deliver the project on time or early completion even in the face of great adversity. Currie and Brown (2004) viewed that the main benefits of the RC procurement route are: (1) reduction in project duration; (2) risk sharing between parties; (3) improved understanding of business objectives; (4) alignment of goals; (5) reduction in management resources; (6) increased scope for innovation; (7) reduction in commercial resources and contractual conflicts; (8) improved long-term relationships; (9) enhanced project interfaces; (10) reduction in documentation; (11) proactive safety regime; and (12) proactive quality regime.

Better working relationships

RC places much importance on the continuous relationships amongst the contracting parties for the success of a project (Rahman and Kumaraswamy, 2005). Ling et al. (2006) opined that RC principles may be mobilized to offer contractual incentives, improved relationships amongst contracting parties, and lubricate any transactional frictions. Rowlinson and Cheung (2004c) viewed that when the protective barrier of paper warfare is broken down by a collaborative approach, then there is no need to formally document every discussion. Direct discussion between parties is legitimized, as is rapid decision making. The consequence is that participants are more comfortable at devolving decision making to appropriate levels within the organization and greater job satisfaction ensues. The close working relationship among owners, constructors, and engineers provides a better working environment of the project (Cook and Hancher, 1990; Bayliss, 2000; Drexler and Larson, 2000; Chan et al., 2003a). Enhanced communication and cooperation, the identification of shared goals and objectives, recognition that problems arise, and the agreement to address problems through a special design procedure facilitates a harmonious working relationship between parties (Construction Industry Institute, 1991; Harback et al., 1994; Matthews et al., 1996; Bayramoglu, 2001). The focus on medium- to long-term relationships is held by MacBeth and Ferguson (1994) to be a major partnering benefit, and it is believed that it compresses the normal learning curve and thereby reduces the normal costs of developing and supporting productive relationships between the parties.

Non-adversarial attitude

The traditional adversarial working relationship between owners and contractors is stressful and inefficient (Cowan et al., 1992). The aim of partnering is to reduce adversarial working relationship that will allow more focus on mutual goals to the benefit of both parties (Construction Industry Institute, 1991; Albanese, 1994; Construction Industry Institute, 1996; Nielsen, 1996; Conley and Gregory, 1999; Gransberg et al., 1999; Black et al., 2000; Chan et al., 2003a). The transforming adversarial working relationship is the actual change mechanism that transfers usual business into a trust-based relationship (Lazar, 1997; Drexler and Larson, 2000).

High level of mutual trust

Partnering recognizes an implied covenant of good-faith dealing by all parties involved (Harback et al., 1994; Lazar, 1997; Green, 1999). Within this atmosphere of cooperation and mutual trust, the parties can jointly determine and evaluate the design, engineering and construction approaches. These usually result in improving cost and schedule performance (Cook and Hancher, 1990; Construction Industry Institute, 1991).

Greater responsiveness to problems

Partnering helps to actualize the delegation of authority or empowerment to the project personnel. The flexibility and responsiveness of owners increases under the partnering agreement (Cook and Hancher, 1990). It helps to ensure problem solving at the lowest possible level of authority. Partners can be more responsive to the short-term emergency, changing project or business needs (Construction Industry Institute, 1991; Green, 1999).

Lower administrative costs

Partnering may provide an effective way to lower administrative cost by eliminating the defensive case building (Abudayyeh, 1994; Construction Industry Institute, 1996; Hellard, 1996b; Black et al., 2000; Chan et al., 2003a). Moreover, the cost to negotiate and administer contracts is decreased as partners become knowledgeable of the counterpart's legal and litigation concerns (Construction Industry Institute, 1991). Hellard (1996b) and Cowan et al. (1992) suggested other interesting benefits that partnering reduced paperwork and simplified administrative procedure. Bayliss (2000) advocated that less paperwork and more face-to-face discussions were made possible in partnering projects.

Increased satisfaction

Rowlinson and Cheung (2004c) stated that more pleasure to go to work was a commonly cited view of RC. When the confrontational nature of the conventional contract is replaced by collaborative and proactive working environment, participants find work more rewarding and enjoyable. In fact, people enjoy working in an atmosphere which allows them to make a positive contribution to moving the project forward. Partnering provides a more conducive environment of achieving project objectives (Matthews et al., 1996). All parties involved will gain benefits from the partnering agreement. Since the customer is closer to the construction process and better informed, partnering enhances customer satisfaction (Nielsen, 1996). Contractors obtain a reasonable profit and are assured of continued work at predetermined profit margins (Moore et al., 1992; Back and Sanders, 1996). Joint satisfaction of all shareholders is possible (Cowan et al., 1992). The work becomes enjoyable rather than a burden or an unreasonable risk (Bates, 1994). The project team comprising the contractors, suppliers, designers have higher levels of satisfaction and necessary actions are taken much sooner based on their active input (Nielsen, 1996). Ross (2003) stated that alliancing can assist in increasing job satisfaction for staff with associated benefits to overall organizational culture.

Reduced litigation

Litigation is a major problem in most construction projects. It does not help to realize potential saving. Under a partnering arrangement, the problems of disputes, claims or litigation are greatly reduced through open communication and improved working relationships (Cook and Hancher, 1990; Construction

Industry Institute, 1991; Abudayyeh, 1994; Construction Industry Institute, 1996). Gransberg et al. (1999) advocated that dispute and claim cost on partnering projects was relatively low. A similar conclusion can be found in the research of Li et al. (2000) and Ruff et al. (1996). The US Army Corps of Engineers and the UK oil industry had used partnering on large and small contracts since 1980s. Bayliss (2002) reported that not a single dispute had escalated to litigation in these partnered projects. This is in sharp contrast to the number of disputes received on non-partnered contracts of similar scale (Schultzel, 1996; Bloom, 1997).

Efficient problem solving

Partnering provides an effective way to develop a control and resolution mechanism for dealing with construction problems (Cowan et al., 1992; Woodrich, 1993; Pena-Mora and Harpoth, 2001; Chan et al., 2003a). The partners recognize potential problems and devise an action plan to address how problems are jointly identified and resolved. The partnering agreement provides each party an opportunity to learn and use the others' problem-solution methods (Cook and Hancher, 1990; Bates, 1994; Hellard, 1996b; Stephenson, 1996; Conley and Gregory, 1999). Sanders and Moore (1992) concluded that partnering helped to eliminate many personal conflicts.

Enhanced communication

RC is based on recognition of mutual benefits and win-win scenarios through more cooperative relationships between parties (Alsagoff and McDermott, 1994; Jones, 2000). To break the traditional hierarchical communication channels, partnering promotes openness, trust and efficient communication through common language (Construction Industry Institute, 1991; Sanders and Moore, 1992; Li et al., 2001; Chan et al., 2003a). Increased communications on various subjects means that it is less likely for parties to be surprised of schedule delays and additional costs, which often lead to disputes and litigation (Sanders and Moore, 1992; Brown, 1994; Moore, 2000; Li et al., 2001). Ross (2003) perceived that alliancing is beneficial to significant increase in communication and general project management skills, thus strengthening the relationship between different parties.

Better safety performance

Taking joint responsibility to ensure a safe working environment for all parties reduces the risk of hazardous working conditions and avoids workplace accidents (Matthews and Rowlinson, 1999). Actually, the safety performance can be improved as partners better understand each other and as the knowledge of construction process and systems improves drastically

(Construction Industry Institute, 1991; Moore et al., 1992; Albanese, 1994; Bates, 1994; Schultzel, 1996). Ross (2003) opined that alliancing is contributory to develop better practice management of health and safety, environment and community.

Improved culture

Bloom (1997) indicated that evaluations of Army partnering contracts had shown a distinct improvement in the culture of the people working on the contract. When people work in a conflict-free environment, they concentrate on the job rather than on potential claims, and the morale and effectiveness of the whole 'team' is improved (Brown, 1994; Ruff et al., 1996; Bloom, 1997). Carrillo (1996) believed that JV may help to reduce cultural problems.

Continuous improvement

RC encourages long-term provisions and mutual planning, and introduces a degree of flexibility into the contract by considering a contract to be a relationship amongst the parties (Macneil, 1974, 1980). The traditional responsibility for improvement is mainly rested on contractors who are assumed to create burdens, while clients and consultants act as sceptical judges (Cowan et al., 1992; Wakeman, 1997). Partnering provides a way for all parties to develop continuous improvement. It is a joint effort and with a long-term focus on eliminating barriers to improvement (Construction Industry Institute, 1991; Ellison and Miller, 1995; Green, 1999; Black et al., 2000; Chan et al, 2003a).

Enhancement of reputation

Ross (2003) reckoned that alliancing is beneficial to build up and enhance the reputations of owner, contractors and other stakeholders. As a result, reputation enhancement can lead to increased prospects of repeated and referred works. Carrillo (1996) opined that JV is contributory to improved track record of a company, thus enhancing its reputation.

Reduction of public financing

To a government, PPP frees up fiscal funds for the needed areas of public service, thus helping to improve cash flow management because high up-front capital expenditure is replaced by periodic service payments and cost certainty (Asian Development Bank, 2007). The total life cycle cost of procured service of facilities is often less than provided by direct government funding. For developing nations, the provision of needed infrastructure and services is enabled where public finance is seriously constrained (Asian Development Bank, 2007).

Drivers for the economy

Carrillo (1996) viewed that JV provides more parties an opportunity to work on large projects, thus providing a strong driver for economic growth. Asian Development Bank (2007) mentioned that PPP provides a path for the private sector participants to enter into the public sector markets. If the price is set accurately and the cost is managed effectively, the projects can provide good profits and investment returns on a long-term basis. As a result, the increased business activities can stimulate economic growth, enhance competitiveness of local firms, and nurture entrepreneurship.

Chapter summary

This chapter, based on a comprehensive literature review on different forms of RC, identifies that there are seventeen major benefits of adopting RC. These include: (1) better cost control; (2) sharing of risk; (3) better quality product; (4) potential for innovation; (5) better time control; (6) better working relationships; (7) lower administrative costs; (8) increased satisfaction; (9) reduced litigation; (10) efficient problem solving; (11) enhanced communication; (12) better safety performance; (13) improved culture; (14) continuous improvement; (15) enhancement of reputation; (16) reduction of public financing; and (17) drivers for the economy. Previous related research works on investigating the benefits of adopting different forms of RC proved that the aforesaid major benefits of RC are valid. In Chapter Five, difficulties in implementing RC will be discussed.

Chapter 5

Difficulties or potential obstacles in implementing relational contracting

Introduction

Different parties working on a project are required to commit themselves when RC is implemented (Palaneeswaran et al., 2003). It is not easy to create trust amongst project team members, and it may be difficult to work through traditional confrontational working attitudes. A major reason for a failure to adopt RC is that different parties only give 'lip service' when implementing this procurement approach (Construction Industry Institute, 1996; Slater, 1998; Cheng et al., 2000; Ross, 2003). In fact, there always exists a win-lose climate when the traditional procurement method is adopted. Therefore, it is very difficult to change to cooperative and collaborative thinking (Larson and Drexler, 1997). To implement RC successfully, the project participants are required to commit themselves to change and work in an integrated team environment that develops win-win harmonious working relationships (Conley and Gregory, 1999; Lazar, 2000).

Research on difficulties or potential obstacles in implementing relational contracting

Partnering in construction: critical study of problems for implementation

Chan et al. (2003b) conducted a comprehensive literature review on the potential obstacles to successful implementation of construction partnering in general and then identified the perceived major problems related to partnering in the Hong Kong construction industry. The empirical findings indicated that the top ten barriers of partnering success were: (1) parties were faced with commercial pressure that compromised partnering attitude; (2) parties had little experience with the partnering approach; (3) uneven levels of commitment were found amongst project participants; (4) risks or rewards were not shared fairly between parties; (5) partnering concept was not fully understood by project participants; (6) dealing with large bureaucratic organizations impeded the partnering effectiveness; (7) conflicts arose from misalignment of personal goals with the project goals; (8) participants were conditioned in a win-lose environment; (9) parties did not have proper training on partnering approach; and (10) partnering relationship created a strong dependency on other partners. It is believed that the empirical results could generate some recommendations to alleviate the potential obstacles to partnering success in future construction projects, such as the identification of critical success factors for construction partnering projects.

Project alliancing – a strategy for avoiding and overcoming adversity

Ross (2001) viewed that there are significant downsides (potential problems) and risks associated with alliancing and they are summarized as follows:

1 Although cost over-runs are shared with the non-owner participants up to the point where the non-owners have lost their limb-2 fee, beyond this point the risk rests wholly on the owner, which has to be capable of carrying the risk on its own balance sheet.
2 It requires significant involvement and commitment of owner personnel and top management to support the alliance process.
3 It requires a great cultural shift from the traditional confrontational working relationship to integration, collaboration, and high performance teamwork. To do so, it involves high costs to establish the alliance and develop and maintain the alliance culture.
4 There may be potential probity issues for public sector projects.
5 It relies heavily on developing and maintaining strong personal and corporate relationships. If they fail, it will create serious problems.
6 It is more difficult for certain insurances to procure for an alliance.

Application of public-private partnerships in urban rail-based transportation project

Asian Development Bank (2007) conducted a research study on the application of PPP in urban rail-based transportation project. After conducting a comprehensive literature review and some case studies of PPP projects, fourteen potential major obstacles to successful implementation of PPP were identified, which included: (1) misallocation of risk; (2) private sector failure; (3) falling service quality; (4) high costs and lengthy time; (5) political and social obstacles; (6) lack of well-established legal framework; and (7) non-conducive financial market. In addition to these, lack of suitable skills and experience, lack of innovations in design and difficulties in seeking financial partners can also affect the implementation of the RC approach.

Technology transfer on joint venture projects in developing countries

Carrillo (1996) conducted a research study on technology transfer on JV projects in developing nations. After interviewing eight UK contractors, he identified eight disadvantages for partners from developed nations to implement JV. These included: (1) ability to reduce potential competition; (2) unstable agreements for a limited period; (3) reduction of competitive advantage; (4) local partner may continually desire more; (5) additional risk; (6) loss of control; (7) diluted earnings; and (8) possible conflict. On the other hand, there were four disadvantages identified for partners from developing countries to adopt JV, which encompassed: (1) may be considered as necessity only; (2) loss of control; (3) work goes to overseas company; and (4) possible dilution of 'single point of responsibility' principle.

Table 5.1 indicates the summary of previous research on investigating the major difficulties/potential obstacles in implementing different forms of RC. The results show that risk sharing failure was the most frequently cited major difficulty for various forms of RC, with conflicts arising from misalignment of personal goals with project goals being the second; and inadequate experience with RC approach, relationship/communication problems, and substantial cost to establish alliance being the third.

Difficulties/potential obstacles in implementing relational contracting

A comprehensive review of the literature shows that the major difficulties/potential obstacles in implementing RC can be grouped under seventeen categories, namely, risk sharing failure, relationship problems, uneven level of commitment, substantial costs to establish the alliance, disreputable relationship, misunderstanding of the RC concept, cultural barriers, communication problems, lack of continuous improvement, inefficient problem solving, inadequate efforts to keep RC going, political and social barriers, professional indemnity insurance, failure of private sector, falling service quality, lack of well-established legal framework, and non-conducive financial market. Figure 5.1 shows the identified major difficulties/potential obstacles in implementing different forms of RC.

Risk sharing failure

Risk sharing failure is a major barrier to the success of a RC project. It may be difficult for project participants to share the risk fairly in the partnering process (Cook and Hancher, 1990; Construction Industry Institute, 1991) because they may be inclined to take full advantage of the partnering spirit to lower their own risk. As a result, parties may be reluctant to share the risk

Table 5.1 Summary of research investigating the major difficulties/potential obstacles in implementing different forms of RC

Major difficulties/potential obstacles in implementing different forms of RC	Partnering (Chan et al., 2003b)	Alliancing (Ross, 2001)	PPP (Asian Development Bank, 2007)	JV (Carillo, 1996)	Total
1. Risk sharing failure	√	√	√	√	4
2. Conflicts arising from misalignment of personal goals with the project goals	√		√	√	3
3. Inadequate experience with the RC approach	√		√		2
4. Relationship/communication problems	√	√			2
5. Substantial costs to establish alliance		√	√		2
6. Commercial pressure to compromise RC attitude	√				1
7. Uneven levels of commitment among project participants	√				1
8. Misunderstanding of RC concept	√				1
9. Dealing with large bureaucratic organizations impeding the RC effectiveness	√				1
10. Win-lose environment among project participants	√				1
11. Inadequate training on RC concept	√				1
12. Cultural problems		√			1
13. Potential probity issues		√			1
14. Professional indemnity insurance		√			1
15. Private sector failure			√		1
16. Falling service quality			√		1
17. Political and social obstacles			√		1
18. Lack of well-established legal framework			√		1
20. Non-conducive financial market			√		1
21. Ability to reduce potential competition				√	1
22. Losing control				√	1
Total	10	6	9	4	

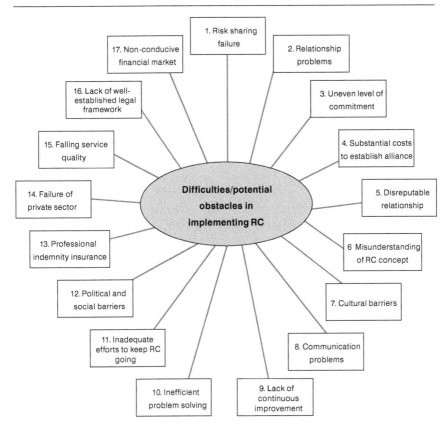

Figure 5.1 Summary of the major difficulties/potential obstacles in implementing different forms of RC

and maintain the working relationship (Larson and Drexler, 1997). Ross (2003) stated that cost over-runs are shared with the non-owner participants up to the point where the non-owners have lost their limb-2 fee. However, beyond this point, the risk rests entirely with the owner and it has to be capable of carrying the risk on its own balance sheet. This factor forms an obstacle to the use of alliancing on financed projects where the financiers generally insist that the risk of cost over-run is transferred off the owner's balance sheet onto others. Asian Development Bank (2007) viewed that if risks are inequitably or wrongly allocated to the party unable to manage or control them, PPP projects would fail and the consequential losses are beyond the capacity of the parties concerned. It may be caused by the government who is too entrenched to transfer risks and by private sector participants who are inexperienced or over-aggressive in taking up risks that are beyond their control. Carrillo (1996) viewed that there may exist additional risks in implementing a JV project.

Relationship problems

Partnering is meant to encourage project team members to alter from their traditional confrontational working attitude to a more cooperative, collaborative and team-based approach to avoid dispute attitude (Moore et al., 1992; Loraine, 1994). The traditional adversarial working relationship always hinders the development of good working relationships between various contracting parties. Carrillo (1996) also opined that it is possible to induce conflicts if a JV project is managed improperly.

Confrontational relationships

Win-win philosophy is a key to partnering success (Hellard, 1996b; Ruff et al., 1996; Lazar, 2000). Nevertheless, a number of parties do not have faith in other parties because of past unpleasant experiences and fear for the unknown and changes (Larson, 1995; Larson and Drexler, 1997). In fact, it is quite difficult for project parties to change the myopic thinking, and they always try to procure benefits out of their relationships and end up with a lose-lose situation (Construction Industry Institute, 1996; Hellard, 1996b).

Distrust

RC does not always work without risk. Although the development of trust amongst different parties is a vital element for partnering success, it may be a high risk in the process of developing it (Cowan et al., 1992). As a matter of fact, the project environment beneficial to developing trust may be affected by the bitter experience in litigation, dispute and past confrontational working relationship (Albanese, 1994; Harback et al., 1994; Lazar, 1997). It is thus difficult to build up trust since parties bring adversarial experiences to partnering. And it is also a barrier in adopting more progressive approaches (Dozzi et al., 1996).

Over-dependency on others

The partnering concept is inclined to emphasize the strengths of partners and, as a result, cannot recompense fundamental weaknesses in the project participants. Partnering sometimes creates strong dependency on other partners (Cook and Hancher, 1990; Construction Industry Institute, 1991). Gardiner and Simmon (1998) also viewed that partnering could be used to strengthen the working relationship and increase trust between parties, but it might be difficult to remove or reduce interdependencies. Thus, project team members should better understand both the benefits of adopting this procurement method and the consequences of non-conformance to project requirements. Partnering is also regarded as a departure from 'business as usual' and individuals and whole organizations usually have feelings of losing control, being at risk or being awkwardly dependent on others (Woodrich, 1993; Bresnen and Marshall, 2000c).

Uneven levels of commitment

RC requires the commitment of all parties. It means that different parties need to overcome the perceived risks of trust and they require actual commitment rather than lip service. Project team members must have total commitment to the RC process. Nevertheless, uneven level of commitment is common in practice because of various goals amongst parties (Moore et al., 1992). Some researchers opined that the lack of agreement of commitment is a major barrier to partnering, and that it is difficult to obtain and maintain commitment (Construction Industry Institute, 1991; Gardiner and Simmon, 1998). As a result, the project may be full of misunderstandings and intractable conflicts. All contracting parties should therefore devote more efforts to balance the levels of commitment on each side (Moore et al., 1992). Ross (2003) stated that a downside for implementing alliancing is that it requires significant involvement and commitment of owner personnel and senior management to support the process, which also implies the uneven level of commitment shared between parties.

Substantial costs to establish alliance

Ross (2003) reckoned that it requires a significant cultural shift to adopt an alliancing approach. This means there must be a great change from the traditional adversarial approach to integration, collaboration and high-performance teamwork. To do so, it involves substantial costs to establish the alliance and develop and maintain the alliance culture. Asian Development Bank (2007) opined that the arrangements of PPP projects are complicated and often involve a number of parties with different conflicting objectives and interests. As a result, PPP projects often need extensive expert input and high front-end costs and take time in deal negotiation. However, these may be unacceptable to all parties and as a result the deal may not materialize in the beginning or may falter in the end.

Disreputable relationship

There are a number of benefits for implementing RC in the construction industry. Examples include better working relationships, open communication, trust, better project performance, innovation and reduction of unnecessary paper work. However, it can induce corruption easily. Currie and Brown (2004) also stated that in government projects, it is easy to raise potential probity issues which have to be managed carefully. To eliminate the discreditable relationship amongst project participants, Harback et al. (1994) suggested that the project participants should develop a comfort zone. In other words, the project participants should avoid establishing too-close relationships to prevent possible allegations of corruption.

Misunderstanding of the RC concept

To make RC a success, it is essential to have a thorough knowledge and understanding of the RC process. In contrast, misunderstanding of the RC concept may be a major problem for RC implementation. Some project participants fail to understand how the partnering relationship can provide a competitive advantage (Cook and Hancher, 1990; Construction Industry Institute, 1991; Construction Industry Institute, 1996). Larson and Drexler (1997) advocated that inadequate experience in the partnering approach affected the understanding and knowledge of project participants. The fair-profit motive was also not fully understood and supported by the project participants because the concept of partnering was not thoroughly understood. Unfamiliarity of the partnering concept by the project participants can cause a failure in partnering (Sanders and Moore, 1992; Harback et al., 1994). Yeung et al. (2007a) viewed that researchers and industrial practitioners are often confused with the definitions of partnering and alliancing, the two major forms of RC, thus resulting unequal foundation for their discussions and comparisons.

Cultural barriers

It is commonly accepted that partnering is opposed to the traditional adversarial working culture of implementing construction projects. As a matter of fact, it is very hard to change established construction culture (Hellard, 1996b; Lazar, 1997). A number of organizations are unwilling to enter into integrated and harmonious working cultures. It is in fact bureaucratic organizations that always impede the effectiveness of partnering (Larson and Drexler, 1997) and project participants are inclined to comprise their partnering attitude when they are faced with commercial pressure.

Communication problems

Communication should be mutual, clear, effective, efficient and open so as to enhance the understanding of the client's requirements. It is believed that partnering can provide a timely, open and direct line of communication amongst all parties. Problems need to be resolved on-site whenever possible (Moore et al., 1992; Sanders and Moore, 1992). Nevertheless, different partners do not normally trust each other fully and are reluctant to communicate and exchange information freely because of the traditional adversarial working culture (Larson and Drexler, 1997). Communication sometimes fails and it results in less efficient cooperation and collaboration and unreasonable demands because of the ignorance of other parties (Harback et al., 1994; Construction Industry Institute, 1996; Gardiner and Simmon, 1998).

Lack of continuous improvement

Traditionally, it is contractors who normally bear the responsibility for continuous improvement. However, the continuous improvement should be a joint effort between clients and contractors to eliminate waste and barriers (Moore et al., 1992; Brown, 1994). As a matter of fact, it is quite difficult to main continuous improvement. Approval time and development costs are two frequent barriers encountered in ongoing improvement efforts. Nevertheless, joint work on improvement schemes yields good recognition of inherent risks of alternative schemes (Cowan et al., 1992).

Inefficient problem solving

Although issues and problems can be lessened in the RC process, it is still possible to create conflicts among project participants (Sanders and Moore, 1992; Brown, 1994). Albanese (1994) agreed that even if the partnering team is willing to identify, confront and resolve problems, many problems still happen because of inefficient problem solving.

Inadequate efforts to keep relational contracting going

To establish the formation of the partnering arrangement requires extra staff, time and resources. It is, however, very costly and inefficient to align all organizations with every partnering undertaking (Larson and Drexler, 1997). In addition, partnering needs nourishment throughout the life cycle of a project. After implementing an initial workshop, it is easy to get back to daily activities and disregard the partnering concept (Moore et al., 1992). In the actual routine operation, project participants often have a number of difficulties in the partnering process which hamper the success of partnering.

Insufficient training

Insufficient training is a hurdle of implementing partnering (Albanese, 1994; Matthews et al., 1996). Construction Industry Institute (1996) explained that inadequate staff training was an essential reason for partnering failure because without a good training background, the project team members do not fully understand the concept of partnering and thus are not easy to implement partnering successfully.

Not Involving key parties

It is misunderstood that partnering should only exist between clients and contractors. Partnering, however, should also involve other parties, encompassing key subcontractors, design consultants and suppliers (Sanders and Moore, 1992; Hinze, 1994; Construction Industry Institute, 1996; Love, 1997; Kumaraswamy and Matthews, 2000). In fact, if they are not involved in the process of partnering, their opinions and advice cannot be sought, thus greatly affecting the overall project performance.

Lack of senior management support

Lack of top management support is a major barrier in initiating partnering. Furthermore, even if top management aggressively pursues the partnering relationship, it is not easy to filter down the concept to project and on-site staff level. Therefore, the partnering concept may be misunderstood by the mid-level staff and the front-line staff (Lazar, 1997). In fact, if the senior management only provides lip service to the partnering approach, there is a little hope to establish successful partnering relationship (Hellard, 1996b) in a construction project.

Political and social barriers

PPP, as a kind of RC project, often incurs political and social issues when it is adopted in public sector projects. Examples include land resumption, town planning, employment, heritage and environmental protection (Asian Development Bank, 2007). More seriously, if it is managed ineffectively, these can induce public opposition, over-blown costs and delays to the projects. Another common complaint by the public is the high fare or tariff charged on the services provided. If the initial project cost is not estimated carefully, or if general economic conditions are beyond expectation, the private sector would face political uphill battles in raising fare or tariff to a level adequate to cover its costs and earn sound profits and returns on investment.

Professional indemnity insurance

A significant drawback to adopting alliancing relates to indemnity insurance for losses due to professional negligence (Ross, 2003; Currie and Brown, 2004). Ross (2003) opined that it is not very flexible for the insurance industry when it comes to amending pre-existing policies to suit the needs of a project alliance. It is particularly problematic to use professional indemnity (PI) because it is necessary to make a legal claim against a design consultant to trigger a conventional PI insurance policy. If an alliance suffers from an internal loss due to professional negligence by one of the alliance participants, the alliance agreement prevents any legal claim between the alliance participants and the conventional PI policy will not react.

Failure of private sector

PPP projects may collapse if the private sector participants fail. To contract out the PPP projects, the government has to ensure that the parties in the private sector consortia are adequately competent and financially capable of taking up the projects. PPP projects will be more complicated to procure and implement if there is a lack of relevant skills and experience of project partners (Asian Development Bank, 2007).

Falling service quality

Without governmental monitoring and overseeing, PPP projects may not achieve the agreed service quality standards. On one hand, it is likely to limit any innovative adaptations by private sector participants under rigidity in planning approvals. The government may insist on setting a number of restrictions and constraints that would bring about increased cost in service delivery. On the other hand, the government can only depend on indirect controls to maintain service standards and to motivate service improvements (Asian Development Bank, 2007). As a matter of fact, the effectiveness of PPP projects much depends on the appropriate level of control by the government.

Lack of a well-established legal framework

Asian Development Bank (2007) stated that PPP projects generally involve a number of legal structures and documents. It is easy to put pressure on the legal system to deal with issues like protection of public interests against the legitimate rights of private sector. Therefore, there should be an independent, equitable and efficient dispute resolution system to deal with disputes arising amongst the PPP parties and to maintain stability of the partnership. If there is a lack of a well-established legal framework, PPP projects incline to failure.

Non-conducive financial market

In general, when implementing public infrastructure projects, private sector investors rely on high levels of debt or capital market instruments to fund the investment. The debt is always either limited-recourse or non-recourse where the lenders need to bear risk. The loan facilities need to be particularly tailored for the project with long-term maturity and financing costs should be suitably priced to reflect the risks involved. The lack of mature financial markets in the host countries is often a barrier to achieving satisfactory completion of PPP deals (Asian Development Bank, 2007).

Chapter summary

After conducting a comprehensive literature review on major difficulties/ potential obstacles in implementing different forms of RC, this chapter has identified seventeen major difficulties/potential obstacles in implementing RC. These include: (1) risk sharing failure; (2) relationship problems; (3) uneven level of commitment; (4) substantial costs to establish the alliance; (5) disreputable relationships; (6) misunderstanding of the RC concept; (7) cultural barriers; (8) communication problems; (9) lack of continuous improvement; (10) inefficient problem solving; (11) inadequate efforts to keep RC going; (12) political and social barriers; (13) professional indemnity insurance; (14) failure of private sector; (15) falling service quality; (16) lack of well-established legal framework; and (17) non-conducive financial market. The research investigating the major difficulties/potential obstacles of implementing different forms of RC proved that the above-mentioned major barriers of RC are valid. In Chapter Six, critical success factors for adopting RC will be discussed.

Chapter 6

Critical success factors for adopting relational contracting

Introduction

Critical success factors (CSFs) are defined as those factors in which success is necessary for each of the major project participants in a project to have the maximum chance of achieving the goals (Smith and Walker, 1994; Tiong et al., 1992; Turner, 2002). In fact, there have been plenty of research studies on CSFs for achieving construction excellence in different forms of RC (including partnering, alliancing, PPP and JV) over the past decade (Associated General Contractors of America, 1991; Construction Industry Institute, 1991; Moore et al., 1992; Sanders and Moore, 1992; Harback et al., 1994, Mohr and Spekman, 1994; Bennett and Jayes, 1995; Larson, 1995; Romancik, 1995; Bellard, 1996; Construction Industry Institute Australia, 1996; Hellard, 1996a; Brooke and Litwin, 1997; Construction Industry Board, 1997; Slater, 1998; Black et al., 2000; Bresnen and Marshall, 2000a; Demirbag and Mirza, 2000; Cheng et al., 2000; Kumaraswamy and Matthews, 2000; Bayramoglu, 2001; Cheng and Li, 2001, 2004a, b; Ng et al., 2002; Chan et al., 2002a, 2004d, 2006; Cheung et al., 2003a; Adnan and Morledge, 2003; Rowlinson and Cheung, 2004c; Gale and Luo, 2004; Beach et al., 2005; Li et al., 2005; Xu et al., 2005; Tang et al., 2006; Asian Development Bank, 2007). This chapter aims to identify all the CSFs for achieving outstanding RC performance in construction by conducting a comprehensive and critical review of literature on success factors for RC projects in construction.

Research on the critical success factors for adopting different forms of relational contracting

Exploring critical success factors for partnering in construction projects

Chan et al. (2004d) reviewed the development of the partnering concept in general and identified CSFs for partnering projects in the Hong Kong construction industry. By using a postal questionnaire survey targeting project participants with hands-on experience in procuring partnering projects, the opinions of different parties, including clients, contractors and consultants, were sought and evaluated in relation to partnering success factors. The relationship between the perception of partnering success and a series of success factors hypothesized in the research study was derived using factor analysis and stepwise multiple regression. The factor analysis shows that the identified 41 independent variables can be grouped under 10 underlying factors: (1) the establishment and communication of a conflict resolution strategy; (2) a willingness to share resources amongst project participants; (3) a clear definition of responsibilities; (4) a commitment to a win-win attitude; (5) regular monitoring of partnering success; (6) mutual trust; (7) willingness to eliminate non-value added activities; (8) early implementation of partnering process; (9) ability to generate innovative ideas; and (10) sub-contractors' involvement. Then, the ten underlying factors were regressed with the perceived partnering performance of construction projects. The results indicated that there were five significant underlying factors for partnering success in Hong Kong, which included: (1) the establishment and communication of a conflict resolution strategy; (2) a willingness to share resources amongst project participants; (3) a clear definition of responsibilities; (4) a commitment to a win-win attitude; and (5) regular monitoring of partnering success. Such an identification of CSFs could help to formulate effective strategies for minimizing construction conflicts and improving project performance.

Forms of collaboration and project delivery in Chinese construction markets: Probable emergence of strategic alliances and design and build

Xu et al. (2005) studied forms of collaboration and project delivery in Chinese construction markets. The authors presented a strengths, weaknesses, opportunities and threats (SWOT) analysis to inform where opportunities for development might be found and to warn of threats to entry. The research also found that mutual trust, synergistic strengths and complementarities, market demand for services, flexibility for both parties and minimum change of top managers were ranked as the top five CSFs for strategic alliances between foreign contractors and design institutes. In addition, medium-sized, state-owned, and large-sized design institutes were ranked to be the first, second and third preferences for strategic partnering by foreign contractors.

Critical success factors for public-private partnerships/private finance initiatives in the UK

Li et al. (2005) conducted an empirical research study on CSFs for PPP/PFI projects in the UK construction industry. By using a postal questionnaire survey, the relative importance of eighteen potential CSFs for PPP/PFI construction projects in the UK was examined. The results indicate that the ten most important factors are: (1) a strong and good private consortium; (2) appropriate risk allocation and risk sharing; (3) available financial market; (4) commitment/responsibility of public/private sectors; (5) thorough and realistic cost-benefit assessment; (6) project technical feasibility; (7) well-organized public agency; (8) good governance; (9) favourable legal framework; and (10) transparency in the procurement process. The authors also conducted factor analysis for the eighteen CSFs and they are grouped into five categories, namely (1) effective procurement; (2) project implementability; (3) government guarantee; (4) favourable economic conditions; and (5) available financial market. The authors expect that these empirical findings should influence policy development towards PPP and the manner in which partners go about the development of PFI projects.

Joint venture projects in the Malaysian construction industry: Factors critical to success

Adnan and Morledge (2003) conducted a comprehensive literature review on CSFs for construction JV projects and 21 factors critical to the success of construction JV projects were identified. Questionnaire surveys were then distributed to industrial practitioners to conduct empirical data analysis. The results show that the top ten CSFs on construction JV projects perceived by Malaysian and foreign contractors were: (1) mutual understanding between parties; (2) inter-partner trust; (3) agreement of contract; (4) commitment; (5) cooperation; (6) financial stability; (7) coordination; (8) communication/ information; (9) management control; and (10) profit.

Table 6.1 indicates the summary of research on investigating the CSFs of adopting different forms of RC. The results show that mutual trust and long-term commitment were the most frequently cited CSFs for various forms of RC, with equitable allocation of risks and top management support being the third.

Critical success factors for relational contracting

A comprehensive literature review on CSFs for construction RC projects shows that they can be grouped under thirty categories, and it should be highlighted that 'mutual trust amongst the project participants'; 'participants willing to provide a long-term commitment to the process'; 'mutual vision, goals and objectives between parties'; 'equitable allocation of risks between parties'; and 'effective and open communication between parties' are the most important CSFs for RC projects in construction. The following sub-sections and Figure 6.1 explain all the CSFs for RC.

Table 6.1 Summary of research investigating the critical success factors (CSFs) for adopting different forms of RC

CSFs for adopting different forms of RC	Partnering (Chan et al., 2004d)	Alliancing (Xu et al., 2005)	PPP (Li et al., 2005)	JV (Adnan and Morledge, 2003)	Total
1. Mutual trust	√	√		√	3
2. Long-term commitment	√		√	√	3
3. Equitable allocation of risks		√	√		2
4. Top management support	√			√	2
5. Willingness to eliminate non-value-added activities	√				1
6. Early implementation of RC process	√				1
7. Ability to generate innovations	√				1
8. Joint problem solution	√				1
9. Adequate and shared resources	√				1
10. Clear definition of responsibilities	√				1
11. Regular monitoring of RC process	√				1
12. Synergistic strengths and complementarities		√			1
13. Market demand for services		√			1
14. Flexibility to change		√			1
15. Minimum change of top managers		√			1
16. A strong private consortium			√		1
17. Available financial market			√		1
18. Thorough and realistic cost/benefit assessment			√		1
19. Adequate legal framework			√		1
20. Local government support			√		1
21. Stable political and social environment			√		1
22. Transparent and efficient procurement			√		1
23. Efficient coordination and cooperation				√	1
24. Effective and open communication				√	1
Total	10	6	9	5	

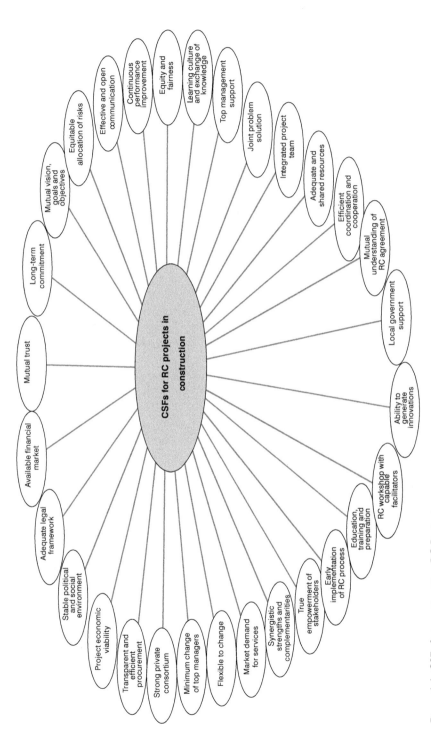

Figure 6.1 CSFs for construction RC Projects

Mutual trust

Many researchers (Construction Industry Institute, 1991; Construction Industry Institute Australia, 1996; Hellard, 1996b; Black et al., 2000; Cheng et al., 2000; Cheng and Li, 2001, 2004a,b; Chan et al., 2002a, 2004d; Beach et al., 2005; Xu et al., 2005; Rowlinson and Cheung, 2004c; and Tang et al., 2006) viewed that mutual trust amongst the project participants is a major success factor for achieving construction excellence in partnering. In fact, when this element is developed, it is easier for the benefits to all parties to be maximized (Construction Industry Institute, 1991; Bennett and Jayes, 1998). Mohr and Spekman (1994) conducted an empirical study and concluded that partnering success is dependent on partnering attributes of commitment, coordination, trust, communication quality and participation, and the conflict resolution technique of joint problem solving. Black et al. (2000) stated that a number of factors, including mutual trust; communication; commitment; a clear understanding of roles; and consistency and a flexible attitude are essential if partnering is to succeed. Cheng et al. (2000) and Bayramoglu (2001) viewed that for partnering to work, parties involved must have mutual trust towards other partners. They should have the belief that others are reliable to fulfil their obligations. It is essential to open the boundaries of the relationship as it can relieve stress and enhance adaptability, information exchange and joint problem solving, and promise better outcomes. Ng et al. (2002) opined that embarking on a project endeavour involves a significant change in attitude from seeking to maximizing individual gains to the continuous search for solutions that benefit all participants. Such relationships begin with respect for other stakeholders, which sees the emergence of trust and teamwork.

Cheung et al. (2003a) examined the behavioural aspect of the participants in construction partnering and found that trust is a critical success factor for partnering. By reviewing a case study, they suggested some tools to build trust, which included contract-specific partnering workshop; partnering charter; monthly partnering review meetings; social functions; and partnering newsletters. Xu et al. (2005) conducted a quantitative analysis and found that mutual trust was ranked first to the success of an alliance project between foreign contractors and design institutes. They stated that the collaborating parties have the right but not the obligation to trust each other under alliance arrangements. However, if mutual trust can be realized, the alliance is likely to be successful. In fact, trust normally comes from both the reputation of the trustee organization and their capacity to adhere to the commitment. The need of trust between partners in a JV project has been identified as a vital element of a long-term joint venture relationship (Madhok, 1995). Adnan and Morledge (2003) agreed that trust is a very important ingredient of managing relationships, and within organizations, trust attributes to more effective implementation of strategy, greater managerial coordination and more effective teamwork. Kwok et al. (2000) stated that it is important for parties to identify and discuss issues broadly, and with mutual trust and commitment to resolve them early. Rowlinson and Cheung (2004c) viewed that trust between RC partners is vital because it creates an opportunity and willingness for further alignment, reduces the need for partners to continually monitor one another's behaviour, reduces the need for formal controls, and reduces the tensions created by short-term inequities. Tang et al. (2006) identified that partnering is a trust-based relationship which is critical to maximizing positive economic outcomes. They pointed out that if trust is present, people can naturally engage in constructive interaction without pondering what hidden motives exchange partners might have, who are formally responsible for problems, or what are the risks in disclosing information. Figure 3.6 illustrates a model of the range of influences that can affect the perception of trust.

Long-term commitment

Long-term commitment is another CSF for achieving outstanding partnering performance as cited by a number of researchers (Construction Industry Institute, 1991; Bennett and Jayes 1995; Romancik, 1995; Construction Industry Institute Australia, 1996; Hellard, 1996; Black et al., 2000; Bresnen and Marshall, 2000a; Cheng et al., 2000; Cheng and Li, 2001, 2004a, b; Chan et al., 2002a, 2004b, d; Rowlinson and Cheung, 2004c; Tang et al., 2006). Long-term commitment can be regarded as the willingness for the involved parties to integrate continuously to the unanticipated problems (Bresnen and Marshall, 2000a; Cheng el al., 2000). More committed parties are expected to balance the attainment of short-term objectives with long-term goals, and achieve both individual and joint missions without raising the fear of opportunistic behaviour (Mohr and Spekman, 1994; Romancik, 1995). All parties should commit their best resources to the PPP project and commitment should be established throughout all management levels (Stonehouse et al., 1996; Kanter, 1999; Li et al., 2005). Adnan and Morledge (2003) opined that commitment reflects the actions of some key decision makers regarding continuation of the relationship, acceptance of the common goals, and the values of the partnership, and the willingness to invest resources in the relationship. Srindharan (1995) showed that commitment is vital because it provides a long-term basis, resources and capabilities to the specific needs of the JV for its success. Kwok et al. (2000) reckoned that the flexibility in resolving on-going problems, the commitment from the JV site management team, the establishment of an efficient and suitable integrated team for the JV operational management will ensure the smooth and successful operation of JV. Black et al. (2000) stated that successful partnering required many factors, in particular a high level of commitment to share goals, preferably including those of the client. Ng et al. (2002) pointed out that stakeholders had to commit to the partnering charter and had to never consider partnering an option unless full commitment was evident. Without the commitment required under the partnering philosophy, it is likely that the direction of the project will suffer from claims, disputation and litigation. Cheng and Li (2002) conducted a quantitative investigation and concluded that the critical functional factors at the stage of partnering reactivation are partnering experience, continuous improvement, learning climate, and long-term commitment. Chan et al. (2004d) explored critical success factors for partnering in construction projects by conducting factor analysis and stepwise multiple regression. They discovered that commitment to win-win attitude was a critical contributor to the personal perception on partnering success. Rowlinson et al. (2006) analysed an alliance project and observed that commitment has a strong impact on the team and alliance culture, indicating that alliance has a high chance of failure when there is an inadequate long-term support from top management. Tang et al. (2006) stated that mutual objectives, active attitude, commitment and equity are the precursors necessary for establishing trust between participants, which is a basis for partnering success. Thorpe and Dugdale (2004) agreed that successful alliance requires commitment by both parties to achieving common goals. Alchimie and Phillips (2003) referred that alliancing is characterized by uncompromising commitments to trust, collaboration, innovation and mutual support to achieve outstanding results. Lendrum (2000) regarded that for the purpose of making partnering successful, all parties have to agree on the objectives and share the principles' process and general information to gain a partner's initial and ongoing support and commitment.

Mutual vision, goals and objectives

Many researchers opined that mutual vision, goals and objectives between parties are essential elements for RC success (Sanders and Moore, 1992; Hellard, 1996a; Bennett and Jayes, 1995; Grant, 1996; Walker et al., 2000a, b; 2002; Adnan and Morledge, 2003). Construction Industry Institute (1991) viewed shared vision (common goals and objectives) as a vital partnering element in which each of the partnered organizations must understand the need for a shared vision and common mission for the partnering relationship. Bennett and Jayes (1995) proposed that one of the three key elements of partnering is to agree common goals to take into account the interests of all the firms involved. Crowley and Karim (1995) agreed that a typical partnering definition is based on its attributes, including trust; shared vision (common goals and objectives); and long-term commitment. Partnering is about people within partnered organizations making a commitment and building trust to work together towards their common project goals and objectives (Walker et al., 2000b; 2002). In fact, at an alliance workshop, the stakeholders identify all respective goals for the project in which their interests overlap. Typically, commonly developed goals include achieving value engineering savings, project delivery on or before time, and maintaining desired quality (Hellard, 1996a; Construction Industry Institute Australia, 1996). Rowlinson and Cheung (2004c) pointed out that RC is defined as a structured management approach to facilitate team working across contractual boundaries. Its fundamental elements include mutual objectives; agreed problem resolution methods; and an active search for continuous measurable improvements. Alliancing can be described as a cooperative arrangement between two or more organizations that forms part of their overall strategy, and contributes to achieving their major common goals and objectives for a specific project (Kwok and Hampson, 1996). Kwok et al. (2000) viewed that the identification of compatible JV partners with strategic objectives which are mutually acceptable and beneficial is a CSF for JV. Thorpe and Dugdale (2004) viewed that alliance contracts are best suited to contracts that require innovation and commitment to achieving common goals. Hauck et al. (2004) also agreed that common goals and objectives are key elements for successful alliance.

Equitable allocation of risks

Gattorna and Walters (1996) argue that a major reason for using a partnering approach to procurement is to develop joint strategies that will achieve strategic objectives. This will enable organizations to improve the return on scarce resources while also reducing their risk. Ng (1997) opined that a major factor in JV for construction projects is the benefit from sharing risk between the parties. With major projects, a single firm might also struggle to raise the financial resources required; through JV, construction firms can combine resources and expertise and thereby increase the likelihood of success. Other researchers, such as Grant (1996), Qiao et al. (2001) and Li et al. (2005), consented that appropriate risk allocation and risk sharing is a key to PPP success. In general terms, this means allocating each risk to the party best able to manage it and this helps to reduce individual risk premiums and the overall cost of the project because the party should be able to manage a particular risk under such the best situation. Carrillo (1996) perceived that JV can assist to spread risks for partners from developed countries and share risks for partners from developing nations. RC enables parties to share the benefits and resources of other parties and develop management and technical advances jointly (Cook and Hancher, 1990; Cowan et al., 1992; Li et al., 2001). It also establishes the tools for both measurement and sharing of gains and risks (Ellison and Miller, 1995; Li et al., 2001). Asian Development Bank (2007) stated that a core principle of PPP arrangement is the allocation of risk to the party best able to manage or control it. Logically, the government would transfer risks associated with asset procurement and service delivery to the private sector participants, who are generally more efficient and experienced in managing them. But the government should be reasonable to take up risks that are outside the control of the private sector participants. Manley and Hampson (2000) showed that one of the alliance features is an equitable risk-reward balance that aligns the commercial interests of the parties.

Effective and open communication

Partnering requires timely communication of information, exchange of ideas, and maintenance of open and direct lines of communication among all project team members (Bellard, 1996; Hellard, 1996a). For construction projects, problems need to be solved at the lowest possible level on-site immediately. If it is only used for routine matters while important issues are conveyed from each site office back to the respective head offices and then back to the site office before any interactions, partnering will fail (Moore et al., 1992; Sanders and Moore, 1992). Larson (1995) reckoned that the key themes and success factors behind partnering are teamwork, collaboration, trust, openness, and mutual respect. Black et al. (2000) conducted a quantitative investigation on factors for partnering success and found that effective communication was rated highly by contractors as a crucial factor for successful partnering. Hellard (1996a) stated that win-win working relationships between partners are achieved by fostering cooperative and mutually beneficial relationships amongst project stakeholders and developing an explicit strategy of commitment and communication. It is clear that effective communication skills can help to facilitate exchange of ideas, visions and overcome difficulties (Cheng et al., 2000). Adnan and Morledge (2003) perceived that without proper communication, problems can occur as a result of the differences between corporate cultures. Cheng and Li (2001) developed a conceptual model of construction partnering to explore the relationship between project and strategic partnering and to determine the critical success factors of partnering. The survey results indicate that open communication is a critical factor affecting both project and strategic partnering. The authors undertook two other research studies and reaffirmed that the four critical common factors affecting the whole partnering process are top management support, open communication, effective coordination and mutual trust (Cheng and Li, 2002, 2004b). In fact, communication between stakeholders is essential whenever an organization is dealing with change and it is equally true when introducing or managing a partnership as communication between parties is vital to understand each party's expectations, attitudes and limitations (Beach et al., 2005). Walker et al. (2000b) stated that alliancing is founded upon team spirit and the honesty associated with notions of trust, commitment, and the application of power and influence. Excellent and effective communication is essential for successful relationship building. Hampson and Kwok (1997) stressed repeatedly that cooperation and communication is a key element of successful alliance. Abrahams and Cullen (1998) opined that cooperative working between entities is an important element for alliance parties to succeed. Hauck et al. (2004) and Walker et al. (2002) pointed out that the intense integration of alliance partners through the whole collaborative process requires excellence in communication at a personal level, at a business level and at an operational level. This generally requires a quantum leap in the use of shared information technology (IT) systems and information processing integration. Tang et al. (2006) agreed that partnering can establish complementary means of communicating information at all levels. Alchimie and Phillips (2003) also agreed that cooperation and collaboration are vital elements for successful alliance.

Continuous performance improvement and evaluation

A key element for partnering and alliancing (Construction Industry Institute, 1991; Bennett and Jayes, 1995, 1998; Walker et al., 2000b) is continuous improvement, meaning that long-term targets are set and achieved by all the stakeholders. RC provides a way for all parties to develop continuous improvement. It is a joint effort with a long-term focus on eliminating wasted barriers to improvement (Construction Industry Institute, 1991; Black et al., 2000; Kumaraswamy and Mattews, 2000; Chan et al., 2004b; Cheng and Li, 2001, 2002, 2004b; Bearch et al., 2005; Li et al., 2005). To implement partnering properly, the stakeholders should agree to a plan for periodic joint evaluation based on the mutual goals to ensure that the plan is proceeding as intended (Hellard, 1996a). Continuous joint evaluation ensures adherence to the agreement and provides a valuable learning process (Construction Industry Institute Australia, 1996). Rowlinson and Cheung (2004c) agreed that a fundamental element for successful alliance encompassed an active search for continuous measurable improvements. Cheng and Li (2004b) also consented that continuous improvement is a vital element for successful partnering to create a good learning culture. Garvin (1993) illustrated that continuous improvement involves continuous learning devoted to gradual process improvement (TQM), radical process improvement (BPR), and learning process improvement (a learning organization) (Kilmann, 1995). Walker et al. (2000b, 2002) observed that an essential element of alliance was continuous improvement in that performance is measured and analysed to provide knowledge about how improvement can be achieved continuously. There must be a commitment to learn from experience and to apply this knowledge to improve performance. Thorpe and Dugdale (2004) addressed that a vital element of alliance was continuous improvement, so as to achieve results on time and to full specification requirements, while innovation will always be required to improve the current process.

Equity and fairness

Equity is another foundation for successful RC implementation. All the interests of stakeholders should be considered in creating mutual goals and there should be commitment to satisfying each stakeholder's requirement based on equity (Construction Industry Institute, 1991; Crowley and Karim, 1995; Li et al., 2000; Kumaraswamy and Matthews, 2000; Walker et al., 2002; Adnan and Morledge, 2003). It reflects a sense of proportionality and balance transcending simple fairness (Construction Industry Institute Australia, 1996). Equity is defined as everyone is rewarded for their work on the basis of fair prices and fair profits (Bennett and Jayes, 1995, 1998). Manley and Hampson (2000) showed that one of the alliance features is an equitable risk-reward balance that aligns the commercial interests of the parties. Hauck et al. (2004) agreed that the foundation of the collaborative process for alliance is equity between parties. Tang et al. (2006) also viewed that equity is one of the precursors essential to establish trust between participants, which is very important for partnering success.

Learning culture and exchange of knowledge

Cheng and Li (2001) compared the critical success factors for project partnering with those for strategic partnering and discovered that as opposed to project partnering, long-term commitment, continuous improvement and learning culture were critical in all three stages of strategic partnering. Another study conducted by the same authors (2002) further analysed that partnering experience, continuous improvement, learning climate and long-term commitment are the critical functional factors at the stage of partnering reactivation. Beach et al. (2005) also pointed out that the free exchange of knowledge with external parties during the partnering process can assist in achieving good partnering performance. Adnan and Morledge (2003) and Qiao et al. (2001) consented that knowledge transfer is a key factor for JV success.

Strong commitment and adequate support from top management

Harback et al. (1994) believed that there are eight essential elements to a successful alliance, which encompasses commitment coming from the top management. It cannot be delegated to a steering committee; it has to come from the personal endorsement of the top leadership of all the players. In fact, strong commitment and support from top management is always a key prerequisite for a successful partnering venture (Bennett and Jayes, 1995; Hellard, 1996a; Slater, 1998; Chan et al., 2002a, 2004b, d; Cheng and Li, 2001, 2002, 2004b). As senior management formulates the strategy and direction of business activities, their full support and commitment are critical in initiating and leading the partnering spirit (Cheng et al., 2000). An empirical study conducted by Black et al. (2000) also proved that commitment from senior management is considered an important factor for partnering success. Beach et al. (2005) consented that management commitment is a critical element of successful partnering, which can assist in achieving outstanding partnering performances.

Productive and joint conflict-resolution mechanism

Since it is very likely to have different goals and expectations amongst different parties, conflicting issues are commonly observed among different project members. Establishment of a productive and joint conflict-resolution mechanism is thus essential to lessen tension between parties (Sanders and Moore, 1992; Bennett and Jayes, 1995; Construction Industry Institute Australia, 1996; Hellard, 1996a; Kumaraswamy and Mattews, 2000; Cheng and Li, 2001; Beach et al., 2005). As a matter of fact, the conflicting parties look for a mutually satisfactory solution and this can be achieved by joint problem solving to seek alternatives for the problematic issues. Such a high level of participation amongst parties may help them to secure a commitment to the mutually agreed solution (Cheng et al., 2000). Adnan and Morledge (2003) reckoned that shared problem solving reflects the degree to which the parties share responsibility both for dealing with problems and maintaining their relationship. Chan et al. (2004d) applied factor analysis and multiple regression analysis to study the relationship between the perception of partnering success and a set of success factors hypothesized in the study. The results indicated that the establishment and communication of a conflict resolution strategy was believed to be the most significant underlying factor for partnering success. Walker et al. (2002) stated that agreed problem resolution is essential when establishing trust and commitment between parties. The CII task force considered that a successful partnering relationship element included conflict resolution through agreed problem solving (Crowley and Karim, 1995). Hampson and Kwok (1997) also proposed that joint problem solving method is a key element of the successful alliances.

Integrated and dedicated project team

Black et al. (2000) conducted an empirical study and identified a dedicated team to be a key to partnering success. Cheng and Li (2001) developed a conceptual model of construction partnering and found that team building was critical at the partnering formation stage of both project and strategic partnering, while it was critical at the application stage of project partnering and the reactivation stage of strategic partnering. Another research study conducted by Cheng and Li (2002) indicated that team building was the most important functional factor with respect to partnering formation, followed by facilitator and partnering agreements. In fact, the use of integrated teams has been said to be crucial to achieving improvements in quality, productivity, health and safety, cash flow, and in reducing project duration and risks (Sanders and Moore, 1992; Beach et al., 2005). Integrated teams can complement the practice of early involvement by facilitating innovation during the design stage of the project, thereby increasing the potential for partnering benefits (Bench et al., 2005). Kumaraswamy and Matthews (2000) and Tang et al. (2006) also viewed that integrated teams are crucial to partnering success.

Adequate resources and willingness to share resources

Owing to the fact that resources are scarce and competitive, it is unusual for an organization to share its own resources with others. Some previous research studies stated that shared resources among parties are important to partnering success (Construction Industry Institute, 1991; Construction Industry Board, 1997; Brooke and Litwin, 1997). The complementary resources from different parties are not only used to strengthen the competitiveness and construction capability of a RC relationship (Cheng et al., 2000), but it is also a major criterion for assessing partnering success. Adnan and Morledge (2003) viewed that to make a JV project a success, the partners have to be willing to cooperate and share information and resources to enable necessary coordination of activity. Chan et al. (2004d) conducted an empirical study and found that willingness to share resources amongst project participants is an underlying success factor for partnering. Cheng and Li (2001, 2002) analysed that adequate resource was one of the critical functional factors at the stage of partnering application.

Efficient coordination and cooperation

Coordination reflects the expectations of each party from the other parties in fulfilling a set of tasks (Mohr and Spekman, 1994). Good coordination which leads to an achievement of stability in an uncertain environment can be attained by an increase in contact points between parties and sharing of project information (Bayramoglu, 2001; Chan et al., 2004b). Cheng and Li (2001) conducted a survey to test a conceptual model of construction partnering and the results indicated that efficient coordination is one of the critical factors affecting partnering. The two authors undertook another research study and identified four common critical success factors for partnering (i.e. top management support, mutual trust, open communication and effective coordination), which were consistent with their previous research study. Adnan and Morledge (2003) opined that cooperation behaviour between companies help to reduce potentially burdensome monitoring and safeguards costs within a JV project.

Mutual understanding of a relational contracting agreement

Xu et al. (2005) viewed that it is important for parties to understand the alliance agreement if the alliance is to succeed. Senior and middle management are usually familiar with the agreement. However, to facilitate the collaboration, all levels of staff from the parties should understand the alliance agreement. William and Lilley (1993) viewed that JV is a means of creating strengths by the partners complementing each other and with proper matching, and it is better for both partners to perceive a vested interest in the success of a JV. Adnan and Morledge (2003) stated that it is vital to maintain the friendly personal contact between the leaders of the cooperating organizations. The careful selection of people who are to work in an alliance will also assist the prospects of mutual trust. In fact, it is better for them to be selected not only based on technical competence but also on an assessment of their ability to form good relationships with people from other organizational and national cultures.

Local government support

Xu et al. (2005) opined that local government support would enhance the alliance because this can assist in creating harmony between foreign contractors and local government and communities. Asian Development Bank (2007) stated that under PPP contracts, the government should be concerned that the assets are procured and services are delivered on time with good quality and meet the agreed service benchmarks throughout the life of the contract. The government should be given the primary role of industry and service regulation. It should also be flexible in adopting innovations and new technology, and provide strong support and make incentive payments to the private sector where appropriate.

Ability to generate innovations

Based on a review of the partnering literature within the management discipline, Cheng et al. (2000) developed a partnering framework to identify the critical success factors for construction parties implementing partnering arrangements. Creativity is one of the critical success factors identified in the framework. The authors stated that creativity has become a common theme in partnering as it could encourage innovative works and management practices. Beach et al. (2005) identified four tools, including (1) integrated teams; (2) partnering workshops; (3) innovative approaches; and (4) dispute resolution planning to build up good partnering relationship between parties. Chan et al. (2004d) applied factor analysis to identify underlying success factors for partnering and one of them is 'ability to generate innovative ideas'.

Implementation of relational contracting workshop with capable facilitators

Facilitated workshops with capable facilitators are also key elements for successful partnering but with relatively less importance (Hellard, 1996a). Green (1999) considered that partnering workshops need to be continuous and not one-off at the project start. Walker et al. (2000b, 2002) pointed out that the interviewing process to derive a shortlist of potential alliancing members requires sophistication and judgement of a client as does the facilitated workshops. This means that alliance workshops are useful tools to help select capable construction alliance partners.

Education, training and preparation

People within a partnered organization need to be well educated so as to be prepared for the operational and cultural change that partnering can entail (Hellard, 1996a; Beach et al., 2005). In fact, good education, training and preparation play a crucial role in determining employee commitment to the concept and practice of partnering and the development of an organizational culture that freely shares knowledge with external partnered organizations (Beach et al., 2005).

Early implementation and regular monitoring of the relational contracting process

Chan et al. (2004d) applied principal component analysis to identify underlying partnering success factors, which included regular monitoring and early implementation of partnering process. They stated that to ensure the success of partnering, multi-project agreements and a list of partner-selection criteria should be developed and started at the design stage of a construction project. Monitoring methods encompass the evaluation of team performance, well-defined roles and responsibilities and determining measurable goals of individual responsibilities. The monitoring process could be ensured by a team leader or partnering champion. The same authors (2004b) conducted a comparative study of partnering practices in Hong Kong and reaffirmed that early implementation of partnering process was one of the most critical success factors for achieving outstanding partnering performance amongst the private, public and infrastructure sector. Other researchers (Hellard, 1996a) also opined that it is essential to implement partnering earlier to obtain more partnering benefits.

True empowerment of stakeholders

Chan et al. (2004b) recommended that to implement partnering efficiently and effectively, partnering should be truly empowered by their senior management. In fact, the primary roles of the partnering champions are to foster the partnering concepts and assist those who might slip back to their old ways to maintain the partnering spirit. The appointment of partnering champions is to ensure that partnering principles do not slip out of focus and by regularly monitoring the partnering process. Indeed, the partnering champions act as a steering group to develop and implement the action plans following the partnering workshops. Individual evaluation forms should be completed by the project team regularly and returned to the nominated champions for compilation. The team then reviews the results and takes the necessary corrective measures.

Synergistic strengths and complementarities between parties

Xu et al. (2005) conducted a research study and found that synergistic strengths and complementarities between parties was ranked to be the second CSF for alliancing between foreign contractors and design institutes. The synergistic strengths and complementarities enable the collaborating parties to utilize external resources effectively and therefore reduce the costs. It should be noted that synergistic strengths and complementarities emphasize that the collaborating parties are able to provide value to each other as well as adding value for the client.

Market demand for services

Xu et al. (2005) discovered that market demand for the services provided by a collaboration of foreign contractor and design institute is highly related to the success of the collaboration. The collaboration is business oriented and if the market demand is decreased, it will be difficult to maintain the alliance.

Flexibility to change

Black et al. (2000) stated that if partnering is to succeed, a flexible attitude is necessary. This means that project team members need to be willing to accept changes. For example, the parties should integrate at all levels and freely share information both formally and informally. Xu et al. (2005) viewed that with the flexibility provided by the alliance parties, the collaborating parties can be quick both to reduce external capacity as prices fall and to increase external capacity as prices rise.

Minimum change of top managers

Xu et al. (2005) stated that stability of top managers is one of the conditions for consistency in company policies, which may be a prerequisite for achieving alliance success. As a matter of fact, it is easier to pursue common goals and objectives if the change of top managers between parties is kept to a minimum.

Strong private consortium

A number of researchers have identified a strong private consortium as a CSF for PPP projects (Jefferies et al., 2002; Tiong, 1996; Birnie, 1999; Li et al., 2005). Before contracting out the PPP projects, the government has to ensure that the parties in the private sector consortium are adequately competent and financially capable of taking up the projects (Asian Development Bank, 2007). Therefore, it is recommended for private companies to explore other participants' strengths and weaknesses and, when appropriate, join together to form consortia capable of synergizing and exploiting their individual strengths. Asian Development Bank (2007) also stated that good working relationships among partners is critical because they all bear relevant risks and get mutual benefits from the cooperation.

Transparent and efficient procurement process

To implement PPP effectively, it is essential to have a transparent and efficient procurement process (Jefferies et al., 1992; Kopp, 1997; Gentry and Fenandez, 1997; Li et al., 2005). Asian Development Bank (2007) viewed that a clear project brief and client requirements assist in achieving these in a bidding process. Very often, competitive bidding exclusively on price may not help to look for a strong private consortium and obtain value for money for the public. Thus, it is better for the government to take a long-term view in seeking the right partner.

Project economic viability

Since the major source of revenue to the private sector is sometimes generated from direct tariffs levied on users, there are revenue risks that can go out of control of the private sector (Asian Development Bank, 2007). In addition, there may be unforeseen risks during the course of project life. To ensure project economic viability, it is rational that the private sector would seek government guarantees, joint investment funding or supplemental periodic service payments to cover the project costs and earn reasonable profits and investment returns. Simultaneously, the government ought to take appropriate consideration of private sector profitability requirements so as to have stable arrangements in PPP projects.

Stable political and social environments

A stable political and social environment is a prerequisite for a successful PPP implementation (Qiao et al., 2001; Zhang et al., 1998; Asian Development Bank, 2007). This in turn relies on the stability and capability of the host government. In fact, political and social issues go beyond private sector domain and they should be handled by the government. If unduly victimized, it is legitimate that the private sector participants should be sufficiently compensated. In a balance between public affordability and private profitability, it is not uncommon to see government paying periodic service payments to private sector participants to cover the project costs and permit them reasonable profits and investment returns.

Adequate legal framework

It is believed that an independent, equitable, efficient and favourable legal framework is a vital factor for successful PPP project implementation (Qiao et al., 2001; Jefferies et al., 2002; McCarthy and Tiong, 1991; Akintoye et al., 2001; Asian Development Bank, 2007). Adequate legal resources at reasonable costs should be available to deal with the amount of legal structuring and documentation required. A transparent and stable legal framework can assist in making the contracts and agreements bankable. An adequate dispute resolution system should be able to help the stability of the PPP arrangements.

Available financial market

A CSF for PPP success is that the private contractor/concessionaire can easily access a financial market with the associated benefits of lower financial costs (Qiao et al., 2001; Jefferies et al., 2002; Akintoye et al., 2001; Li et al., 2005). In fact, an accessible financial market is an incentive to private sector interests in taking part in PPP projects. Asian Development Bank (2007) stated that project financing is a critical factor for private sector investment in public infrastructure projects. It would be an incentive for private sectors to take up PPP projects when there is an efficient and mature financial market to benefit low financing costs and diversified range of financial products.

Chapter summary

This chapter has conducted a comprehensive literature review on CSFs for construction RC projects and a conceptual framework has been developed.

A total of 30 CSFs have been sought and discussed within this conceptual framework. Research on the CSFs for different forms of RC proved that the aforesaid CSFs for construction RC projects are sound and valid, therefore, they can be used as a basis for studying the CSF for construction RC in different nations. By doing so, a best global practice framework for RC can be ultimately developed. In Chapter Seven, performance measures for RC in construction will be discussed.

Chapter 7

Conceptual framework for identifying key performance indicators for relational contracting

Introduction

Over the past decade, there have been abundant research studies looking into benefits of adopting RC (Macneil, 1978; Alsagoff and McDermott, 1994; Dozzi et al., 1996; Larson and Drexler, 1997; Walker and Chau, 1999; Jones, 2000; Rowlinson and Cheung, 2004c; Kumaraswamy et al., 2005); critical success factors for achieving construction excellence in RC (Construction Industry Institute, 1991; Moore et al., 1992; Mohr and Spekman, 1994; Construction Industry Board, 1997; Bresnen and Marshall, 2000c; Lazar, 2000; Cheng et al., 2000; Chan et al., 2004b, 2004d; Rowlinson and Cheung, 2004c; Beach et al., 2005; Li et al., 2005; Xu et al., 2005; Tang et al., 2006; Asian Development Bank, 2007); difficulties in implementing RC (Cook and Hancher, 1990; Construction Industry Institute, 1991, 1996; Sanders and Moore, 1992; Albanese, 1994; Larson, 1995; Carrillo, 1996; Larson and Drexler, 1997; Love, 1997; Bresnen and Marshall, 2000c; Chan et al., 2003; Ross, 2003; Asian Development Bank, 2007); and process, conceptual and theoretical models of construction RC (Rahman and Kumaraswamy, 2005; Kumaraswamy et al., 2005) within the construction management discipline. However, unlike other types of building and construction projects, such as hospitals (Chan and Chan, 2004) and design and build projects (Chan et al., 2002b), there is a lack of attention in terms of key performance indicators (KPIs) to comprehensively, systematically and accurately measure the performance of RC projects. As a result, it is difficult for construction senior executives and project managers to evaluate the performance level of their individual RC projects objectively.

This chapter aims to develop a conceptual framework to identify KPIs for measuring the performance of RC projects, first by discussing the definition and functions of KPIs. After that, a comprehensive and critical review of literature on performance measures for RC over the last decade is discussed. A preliminary conceptual framework of performance measures for RC projects was thus formed. To validate the usefulness of the preliminary, conceptual KPI framework for RC projects, an in-depth study of 17 projects from the Hong Kong demonstration projects committee was followed.

The committee was set up in Hong Kong in 2003, and aims to establish a framework, whereby construction industry professionals can collectively set benchmarks for delivering projects and utilize innovative techniques or processes. Ultimately, the goal is to improve the efficiency of the construction process within Hong Kong and Asia as a whole. The committee is composed of about 25 leading practitioners within the Hong Kong construction industry with a number of diverse roles and responsibilities, encompassing property developers, architects, structural and civil engineers, building services engineers, government bodies, main contracting organizations, and subcontracting organizations. After validation, a consolidated, conceptual framework of performance measures for RC was developed.

Research on performance measures for different forms of relational contracting

Yeung et al. (2009) conducted an empirical study using the Delphi survey technique to formulate a model to assess the success of relationship-based construction projects in Australia. Four rounds of Delphi questionnaire surveys were conducted with 22 construction experts in Australia. The results indicated that there are eight KPIs to evaluate the success of relationship-based projects in Australia. These KPIs included: (1) client satisfaction; (2) cost performance; (3) quality performance; (4) time performance; (5) effective communications; (6) safety performance; (7) trust and respect; and (8) innovation and improvement.

To demonstrate the applicability of KPIs and PI to measure the performance of relationship-based construction projects in Australia, three case studies were examined and the data were provided by respondents from another questionnaire survey undertaken by the same research team. Scope of analysis under each case study covers the project performance on client satisfaction, cost, quality, time, communications, safety, trust and respect, together with innovation and improvement. Table 7.1 shows the summary of the background information and the results of different KPI and PI of these three Australian case studies. It should be noted that the PI is applied together with the quantitative indicators (QIs) and the fuzzy quantitative requirements (FQRs) to mitigate the subjectivity problem in interpreting each of the identified KPIs.

Luo (2001) conducted a research study on assessing management and performance of Sino-foreign construction joint ventures in China's construction industry. The study adopted an integrated research approach with a questionnaire survey, supplemented by in-depth interviews to collect both quantitative and qualitative data. It has been found that dominant management control over JVs by Chinese partners is the major form of management, and Sino-foreign construction JVs perform well. In addition, a valid measure, which combines both objective (financial indicators: cost

Table 7.1 Case studies – application of KPIs and PI (reproduced from Yeung et al., 2009 with permission for both print and online use from the *Journal of Management in Engineering*)

	Case 1	Case 2	Case 3
Background			
Nature of project	Building work	Civil work	Civil work
Type of project	Other types of collaborative project	Alliancing project	Alliancing project
Procurement method	Management contracting	Unknown	Unknown
Tendering method	Selective tendering	Negotiated tendering	Unknown
Form of contract	Guaranteed maximum price	Target cost	Unknown
Total contract duration	28 months	15 months	2.5 years
Total contract sum	AUD$100 million	AUD$250 million	AUD$80 million
KPI survey result			
Client's satisfaction Score	9 out of 10 for client satisfaction	9 out of 10 for client satisfaction	8 out of 10 for client satisfaction
Cost performance	Underrun budget by 7.5%	On budget	Under-run budget by 8%
Quality performance	3 – average number of non-conformance reports generated per month	10 – average number of non-conformance reports generated per month	2 – average number of non-conformance reports generated per month
Time performance	On schedule	On schedule	Ahead schedule by 12%
Safety performance	1 (measured in terms of LTIFR 1,000,000 working hours)	1 (measured in terms of LTIFR 1,000,000 working hours)	0 (measured in terms of LTIFR 1,000,000 working hours)
Effective communications	8 out of 10 for effective communications	8 out of 10 for effective communications	9.5 out of 10 for effective communication
Trust and respect	4 weeks (average duration for settling variation orders)	3 weeks (average duration for settling variation orders)	Less than 1 week (average duration for settling variation order)
Innovation and improvement	8 out of 10 for innovation and improvement	8 out of 10 for innovation and improvement	9.5 out of 10 for innovation and improvement
Performance index	3.812 out of 5	3.271 out of 5	4.849 out of 5

and profitability, JV survival and JV duration) and subjective measures (e.g. achievement of main strategic objectives, overall performance assessment by parents or JV management) to assess the JV's performance in terms of its social role in developing nations.

Definition and functions of key performance indicators

Cox et al. (2003) defined KPIs as compilations of data measures used to evaluate the performance of a construction operation. They are the methods that management uses to assess employee performance of a particular task. These evaluations typically compare the actual and estimated performance in terms of efficiency, effectiveness and quality in terms of product and workmanship.

KPIs aim to enable measurement of project and organizational performance throughout the construction industry (The KPI Working Group, 2000). This information can then be used for benchmarking purposes, and will be a key component of any organization's move towards achieving best practice. Collin (2002) stated that the aim of introducing KPIs is to objectively measure a range of fundamental characteristics associated with procurement systems to identify elements or aspects that have changed as a result of amendments to the procurement process as well as establish the strengths and weaknesses of each procurement system. He further added that before a set of KPIs is developed, it is vital to achieve agreement on what broadly constitutes procurement performance. Such an approach is commonly referred as determining the key result areas (KRAs). Once the KRAs are agreed, then measures (KPIs) can be developed to support them. Figure 7.1 shows the relationship between KRAs and KPIs.

Collin (2002) also advocated that the process of developing KPIs involved the consideration of the following eight factors:

1 KPIs are general indicators of performance that focus on critical aspects of outputs or outcomes;
2 Only a limited, manageable number of KPIs is maintainable for regular use. It can be time and resource consuming if there are too many (or too complex) KPIs;
3 It is essential to have a systematic use of KPIs because their value is almost completely derived from their consistent use over a number of projects;
4 Data collection must be made as simple as possible;
5 A large sample size is required to reduce the impact of project-specific variables;

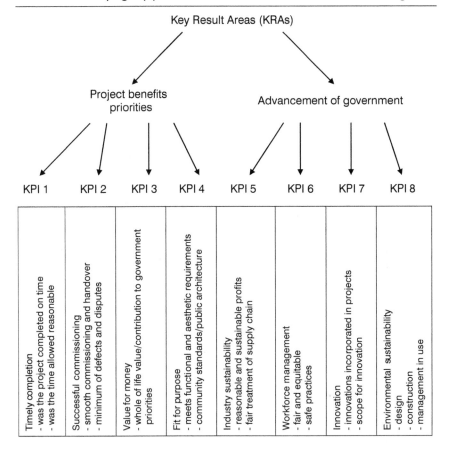

Figure 7.1 Relationship between KRAs and KPIs (measurements of project success) (adapted from Collin, 2002)

6 For performance measurement to be effective, the measures or indicators must be accepted, understood and owned across the organization;

7 KPIs will need to evolve and it is likely that a series of KPIs will be subject to change and refinement;

8 Graphic displays of KPIs need to be simple in design, easy to be updated and accessible.

With these factors in mind, a preliminary conceptual framework is developed to identify KPIs for measuring the performance of RC projects based on a critical and comprehensive literature review on performance measures for RC projects (including partnering, alliancing, PPP and JV), followed by verification and validation through in-depth study of 17 Hong Kong demonstration projects.

Conceptual measures for assessing relational contracting projects

Before developing a conceptual framework to identify KPIs for RC projects, it is important to understand the inter-relationship between their goals, processes, performances and feedback because it accounts for the need to develop a systematic approach to measure RC performance. Cheung et al. (2003b) proposed that a partnering process can be seen as a system that encompasses four key elements, including goal, process, performance and feedback (Figure 7.2). In any partnering arrangement, the first step is to identify overall project goals, followed by developing strategies that direct efforts to achieve those goals. Then, performance is monitored and measured to evaluate progress. In this sense, measures of partnering performance ought to be reflective of the project goals because each partnering project requires a unique set of measures. Clearly, a partnering project cannot be successful if any one of the four elements is missing. Therefore, measures must be closely related to the project goals, objectives, and strategies. Crane et al. (1999) created a model called 'objectives, goals, strategies, measures' (OGSM) to exemplify the systematic selection of project measures for monitoring partnering performance (Figure 7.3).

Having conducted a comprehensive and critical literature review, 19 performance measures for RC projects were identified, including (1) time performance; (2) cost performance; (3) quality performance; (4) claim occurrence and magnitude; (5) effective communications; (6) safety performance; (7) environmental performance; (8) trust and respect; (9) harmonious working relationships; (10) litigation occurrence and magnitude; (11) dispute occurrence and magnitude; (12) customer satisfaction; (13) profit and financial objectives; (14) scope of rework; (15) productivity; (16)

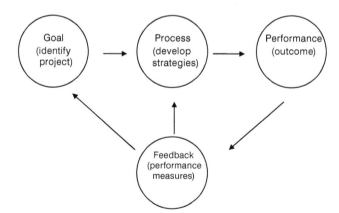

Figure 7.2 Partnering system (reproduced from Cheung et al., 2003 with permission for both print and online use from Elsevier).

Objective	Goal	Strategies	Measures
		Establish key focus areas, with goals, for improvement and tracking	Focus areas established
	$10,000,000 reduction in year 1		
10% reduction in capital spending budget over 4 years			Progress against goals
			Charter established
	$50,000,000 cumulative reduction by end of year 4	Charter each key manager on EPC team with one or more goals in key focus areas	
			Progress against goals
			Steering team established
		Project steering team review progress against goals monthly, and take necessary corrective actions	Effectiveness of monthly reviews

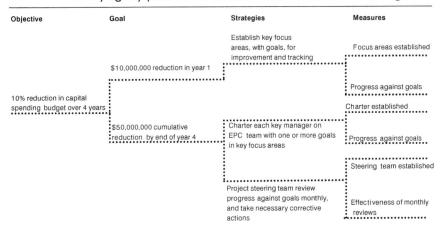

Figure 7.3 OGSM Model (adapted from Crane et al., 1999)

innovation and improvement; (17) pollution occurrence; (18) professional image establishment; and (19) employees' attitude.

After identifying 19 performance measures for RC projects based on a comprehensive and critical literature review, 17 projects from the Hong Kong demonstration projects committee were studied to validate the preliminary KPI conceptual framework for RC projects. Table 7.2 shows the KPIs used for these 17 projects with partnering approach. The results show that safety performance was the most frequently cited KPI, with quality performance and time performance coming second; effective communications and profit and financial objectives came third; environmental performance came fourth; cost and professional image establishment, fifth; innovation, sixth; long-term business relationship, improved working relationship, job satisfaction, and trust and respect, seventh; client satisfaction came eighth; and top management commitment, reduction of paperwork and partnering workshops following.

Clearly, the majority of KPIs used for these partnering projects are identical to the performance measures identified in the literature. However, six performance measures identified in Table 7.2 are not mentioned previously and it is suggested that these additional attributes should be included in the preliminary KPI conceptual framework. These additional KPIs are: (1) long-term business relationship; (2) client satisfaction; (3) job satisfaction; (4) top management commitment; (5) introduction of facilitated workshops; and (6) reduction of rework. In total, a consolidated framework comprising 25 performance measures has been developed (Figure 7.4).

Crane et al. (1999) introduced three types of measures for partnering projects. Two of them, result measures and relationship measures, are vital to reflect the performance of a partnering project. Result measures are 'hard'

Table 7.2 Key performance indicators (KPIs) for 17 demonstration projects using partnering approach in Hong Kong (reproduced from Yeung et al., 2007b) with permission for both print and online use from Taylor & Francis Journals (UK)

	1. Safety	2. Quality/High quality	3. Anticipated delay/Programme	4. Effective communications/Teamwork	5. Profit/Financial objective	6. Environmental performance/Environmental protection	7. Cost/Savings	8. Successful project/Elevated project image/Professional image establishment	9. Innovation	10. Relationships/Improved relationship	11. Team of next project/Long-term relationship	12. Job satisfaction	13. Trust/Respect	14. Client satisfaction	15. Top management commitment	16. Reduction of paperwork	17. Partnering workshop	*Total number of KPIs identified from each project*
The Orchards	✓	✓	✓	✓	✓	✓	✓	✓	✓		✓		✓					11
Three Pacific Palace	✓	✓	✓	✓	✓		✓	✓	✓	✓	✓	✓						11
Cambridge House	✓	✓	✓	✓	✓	✓	✓	✓	✓		✓			✓				10
Po Lam Road Phase 1	✓	✓	✓	✓	✓	✓		✓	✓			✓		✓				10
Tsim Sha Tsui Station Modification Works	✓	✓	✓	✓	✓			✓			✓	✓	✓					9
Chater House	✓	✓			✓	✓	✓			✓			✓	✓			✓	8
Choi Yuen Phase 2	✓	✓	✓	✓	✓	✓		✓		✓								8
Stonecutters Bridge	✓	✓	✓		✓	✓		✓	✓	✓								8
Grand Promenade	✓		✓	✓			✓	✓			✓		✓	✓				7
Hong Kong Museum of Coastal Defence	✓	✓	✓	✓	✓			✓						✓				7
Tradeport Hong Kong Logistics Centre	✓			✓	✓	✓	✓	✓	✓									6
Design and build of improvement to Castle Peak Road	✓			✓		✓	✓			✓								5
Lok Ma Chau Viaduct	✓	✓	✓					✓							✓			5
Tseung Kwan O Technology Park		✓	✓	✓												✓	✓	5
East Hall Extension of Passenger Terminal Building of the Hong Kong International Airport	✓	✓	✓					✓										4
One Peking Road	✓			✓		✓			✓									4
Tseung Kwan O Area 73A Phase 2	✓	✓				✓												3
Total number of hits for the same KPI	16	13	13	11	11	10	9	9	8	5	5	5	5	3	1	1	1	

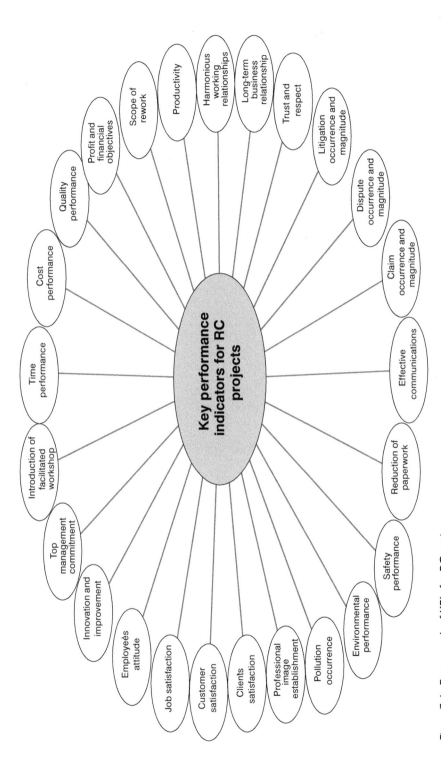

Figure 7.4 Framework of KPIs for RC projects

measures based on performance. Examples cited by Crane et al. (1999) include cost, schedule, quality and safety. Cost and schedule variance adheres to the original estimate and schedule. Quality typically includes such measures as the amount of rework required. Safety can be measured by compiling safety statistics such as lost time incidents. Since result measures tell the decision maker little or nothing about the condition of the environment in which the performance is attained, a partnering relationship must make use of relationship measures to achieve a greater degree of foresight and realize the benefits of increased time to react to problems in the relationship. Cheung et al. (2003b) also pointed out that hard measures alone do not provide a clear picture of partnering performance, as partnering is about cooperative working relationships between parties. They also suggested that it is necessary to use relationship measures to assess the behavioural aspects of partnering. Crane et al. (1999) defined relationship measures as 'soft' measures and they are used to track the activities and effectiveness of the partnered team. The followings are examples of relationship measures:

1   internal communication
2   external communication
3   meeting effectiveness
4   worker morale
5   internal trust/candour
6   external trust/candour
7   internal leadership
8   external leadership
9   accomplishment of objectives
10   utilization of resources
11   problem solving
12   creativity and synergy
13   timely evaluation and appropriate response
14   definition and adherence to roles and responsibilities
15   continuous improvement
16   teamwork

By using a similar approach, the 25 performance measures identified previously can be classified into result-oriented measures and relationship-oriented measures. Result-oriented measures are: (1) time performance; (2) cost performance; (3) profit and financial objectives; (4) quality performance; (5) scope of rework; (6) productivity; (7) safety performance; (8) environmental performance; (9) pollution occurrence; (10) professional image establishment; (11) client satisfaction; (12) customer satisfaction; (13) job satisfaction; and (14) innovation and improvement. Relationship-oriented measures encompass (1) harmonious working relationships; (2) long-term business relationship; (3) trust and respect; (4) effective communications; (5) reduction of paperwork; (6) litigation occurrence and magnitude; (7) dispute occurrence and magnitude; (8) claim occurrence and magnitude; (9) employees' attitude; (10) top management commitment; and (11) introduction of facilitated workshops.

In addition to this classification, measures to reflect the objectives determined by the project team members are always be classified as objective or subjective (Chan, 1996; Chan et al., 2002b; Chan and Chan, 2004). Therefore, the identified 25 performance measures can also be divided into objective and subjective measures. The calculation methods of the proposed KPIs are then divided into two groups. The first group uses mathematical formulae to calculate the respective values. Formulae will be presented after the brief explanation of each KPI, such as time performance, cost performance, safety performance, and environmental issues. The other group uses subjective opinions and personal judgement of the stakeholders. Examples include quality, trust and respect, effective communications, and so on. Figure 7.5 shows the consolidated framework of KPIs for RC projects.

The following sub-sections briefly present all KPIs that constitute the whole framework as they are described in the literature and then verified by the 17 Hong Kong demonstration projects with partnering approach.

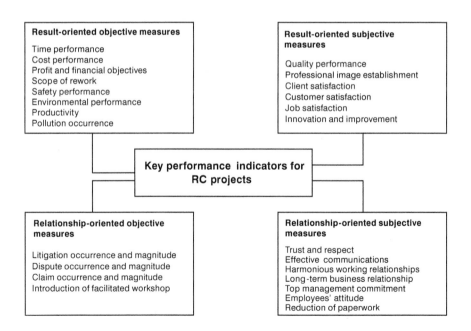

Figure 7.5 Consolidated conceptual framework of KPIs for RC projects (reproduced from Yeung et al., 2009 with permission for both print and online use from the *Journal of Management in Engineering*)

Result-oriented objective measures

Figure 7.5 shows that there are seven result-oriented objective measures, including (1) time performance; (2) cost performance; (3) profit and financial objectives; (4) scope of rework; (5) safety performance; (6) environmental performance; and (7) pollution occurrence.

Time performance

'Time' refers to the duration for completing the project (Chan and Chan, 2004). It is scheduled to enable the building to be used by a date determined by the client's plans (Hatush and Skitmore, 1997). It can be measured by time variation (over-run, on time or under-run) (Naoum, 1994; Chan et al., 2001, 2004b, c, d, 2006; Collin, 2002; Zhao, 2002; Cheung et al., 2003b; Cox et al., 2003; Bayliss, 2004), construction time (Harrigan, 1986; Kogut, 1988; Chan, 1996; Bennett and Jayes, 1998; Crane et al., 1999; Chan et al., 2004b, c, d, 2006), and speed of construction (A1-Meshekeh and Langford, 1999; Chan et al., 2004b, c, d, 2006). Table 7.3 illustrates the definitions of each measurement of time.

Table 7.3 Types of time performance measurement (reproduced from Chan et al., 2002b with permission for both print and online use from the *Journal of Management in Engineering*)

Year	Author(s)	Measurement	Definition
1994	Naoum	Time over-run/Time	The percentage of
2001	Chan et al	under-run	increase or decrease
2002	Collin		in the estimated
2002	Zhao		project in days/weeks,
2003b	Cheung et al		discounting the effect
2003	Cox et al		of extension of time
2003	Martin		(EOT) granted by the
2004b,c,d	Chan et al		client
2004	Bayliss et al		
2006	Chan et al		
1996	Chan	Construction time	Absolute time that
1998	Construction Task Force		is calculated as the number of days/weeks
1999	Crane et al		from start on site to
2004b,c,d	Chan et al		practical completion of
2006	Chan et al		the project
1999	Al-Meshekeh and Langford	Speed of construction	Gross floor area (in square meters) divided
2004b,c,d	Chan et al		by the construction
2006	Chan et al		time (in days)

Cost performance

Cost performance is another vital result-oriented objective measure (Tomlinson, 1970; Good, 1972; Lecraw, 1983; Chan et al., 2006). It is defined as the degree to which the general conditions promote the completion of a project within the estimated budget (Bubshait and Almohawis, 1994). It can be measured by cost over-run/under-run (Crane et al., 1999; Chan et al., 2001, 2004b, c, d, 2006; Zhao, 2002; Cheung et al., 2003b) and unit cost (Chan, 1996; Chan et al., 2002b; Chan and Chan, 2004). Percentage net variation over final cost (%NETVAR) gives an indication of cost over-run or under-run, and it is calculated as the ratio of net variations to final contract sum expressed in percentage term. Table 7.4 shows the definitions of each measurement of cost.

Profit and financial objective

Profit is one of the most vital result-oriented objective measures because most projects, including RC projects, are for profit, and the clients always try to maximize their profit (Geringer and Hebert, 1991; Chan et al., 2004b,c,d; Luo, 2001). Norris (1990) measured profit as the increment by which revenues exceed cost; that is, profit is measured as the total net revenue over total costs. Another common measure of financial achievement is net present value (NPV) (Chan and Chan, 2004). Table 7.2 indicates that 11 out of the 17 Hong Kong demonstration projects using partnering approach use this measure to evaluate their project performance. These projects are (1) The Orchards; (2) Three Pacific Place; (3) Cambridge House; (4) Po Lam Road Phase 1; (5) Tsim Sha Tsui Station Modification Works; (6) Chater House; (7) Choi Yuen Phase 2; (8) Stonecutters Bridge; (9) Hong Kong Museum

Table 7.4 Types of cost performance measurement (reproduced from Chan et al., 2002b with permission for both print and online use from the *Journal of Management in Engineering*)

Year	Author(s)	Measurement	Definition
1994 1999 2001 2002 2003 2004b,c,d 2006	Yeong Crane et al Chan et al Zhao Cheung et al Chan et al Chan et al	Cost over-run/under-run	Increase or decrease in budget (in dollars)
1996 2002b 2004	Chan Chan et al Chan and Chan	Unit cost	Final contract sum (in dollars) divided by gross floor area (in square meters)

of Coastal Defence; (10) Tradeport Hong Kong Logistics Centre; and (11) Design and Build of Improvement to Castle Peak Road.

Scope of rework

Cox et al. (2003) stated that in general, rework takes between 6 and 12 per cent of the total expenditure for a construction project. Nevertheless, the costs arising from rework are at a premium and they sharply increase the total cost of running the project. It is an effective tool to measure overall project performance by calculating the change in the number of man-hours and material costs for repairing works in place or re-handling materials. In fact, when the amount of rework on a job is reduced, both the costs and time associated with the specific task can greatly reduce while the profits dramatically increase.

Safety performance

Safety is defined as the degree to which the general conditions promote the completion of a project without major accidents or injuries (Bubshait and Almohawis, 1994). The issue of safety is of great concern for partnering measures (Crane et al., 1999; Cheung et al., 2003b; Bayliss et al., 2004; Chan et al., 2004b, c, d, 2006). It is a common practice that the measurement of safety is mainly focused on the construction period because most accidents occur during this stage.

Environmental performance

It is well known that the construction industry has an adverse effect on environmental performance. Songer and Molenaar (1997) reported that 14 million tonnes of waste are put into landfill in Australia each year, and 44 per cent of that came from the construction and demolition industry. Three kinds of indicators (Table 7.5), including International Organization for Standardization 14000 (ISO14000) (Chan and Chan, 2004), Environmental Impact Assessment (EIA) score (Environmental Protection Department, 2000), and total number of complaints received caused by environmental issues (Cheung et al., 2003b; Chan and Chan, 2004; Bayliss et al., 2004; Chan et al., 2004b, c, d, 2006), can be used to reflect the environmental performance of a partnering project. Table 7.2 shows that 10 out of 17 Hong Kong demonstration projects using partnering approach adopted environmental performance as a KPI. These projects include (1) The Orchards; (2) Cambridge House; (3) Po Lam Road Phase 1; (4) Chater House; (5) Choi Yuen Phase 2; (6) Stonecutters Bridge; (7) Tradeport Hong Kong Logistics Centre; (8) Design and Build of Improvement to Castle Peak Road; (9) One Peking Road; and (10) Tseung Kwan O Area 73A Phase 2.

Table 7.5 Measures of environmental performance

Year	Author(s)	Measurement
2004	Chan and Chan	ISO14000
2000	Environmental Protection Department	EIA Score
2003	Cheung et al	Total number of complaints
2004	Chan and Chan	received caused by the
2004b,c,d	Chan et al	environmental issues
2004	Bayliss et al	
2006	Chan et al	

Productivity

Productivity is one of the result-oriented objective measures because it is a main key to the cost-effectiveness of projects (Taylor, 1992). Chan (1996) explained productivity as the amount of resource input to complete a given task, and it is often evaluated on a ranked basis. Zhao (2002) measures productivity as number or percentage of collaborative projects finished within time and budget.

Pollution occurrence and magnitude

Complementary to environmental issues, pollution occurrence and magnitude is also a vital result-oriented objective measure because it directly reflects the impact of a construction project on the environment and the society.

Result-oriented subjective measures

Figure 7.5 shows that there are six result-oriented subjective measures, including (1) quality performance; (2) professional image establishment; (3) client satisfaction; (4) customer satisfaction; (5) job satisfaction; and (6) innovation and improvement. These measures are measured subjectively using a seven-point Likert scale ranging from 1 = extremely low level, 2 = low level, 3 = moderately low level, 4 = neutral, 5 = moderately high level, 6 = high level, to 7 = extremely high level.

Quality performance

Quality is an important result-oriented subjective measure for partnering projects that is often cited by researchers (Crane et al., 1999; Chan et al., 2001, 2004b, c, d, 2006; Cheung et al., 2003b; Bayliss et al., 2004). However, different people assess quality differently because it is rather subjective. Quality is defined as the degree to which the general conditions meet the project's established requirements of materials and workmanship (Bubshait and Almohawis, 1994). Crane et al. (1999) used eleven measures to assess quality, including (1) conformance to specifications; (2) achievement of operating objectives; (3) per cent of rework; (4) plant output; (5) participation in design by construction/manufacturing personnel; (6) start-up performance; (7) number of engineering changes; (8) customer feedback; (9) audit deviations; (10) errors and omissions; and (11) first-pass yield. Cheung et al. (2003) defined quality as a measure of how well the work is completed in accordance with the design work. Bayliss et al. (2004) measured quality by counting non-conformance reports and time taken to rectify. Chan et al. (2001) measured quality by the satisfaction level of partnering participants towards the quality of a construction project. Chan et al. (2004b, c, d, 2006) also measured quality performance by using a seven-point Likert scale ranging from very high quality to very low quality. Table 7.6 shows the measures of quality for partnering projects used by previous researchers.

Professional image establishment

Professional image establishment is a vital result-oriented subjective measure (Chan et al., 2004b, c, d, 2006) because it reflects the degree of pride and reputation of each contracting party enhanced by the successful completion of a project. Table 7.2 indicates that 9 out of 17 Hong Kong demonstration projects discussed use this KPI to measure their project performance. These projects are (1) The Orchards; (2) Three Pacific Place; (3) Cambridge House; (4) Po Lam Road Phase 1; (5) Choi Yuen Phase 2; (6) Stonecutters Bridge;

Table 7.6 Measures of quality performance

Year	Author(s)	Measurement
1999	Crane et al	Conformance to specifications Achievement of operating objectives Percent of rework Plant output Participation in design by construction/manufacturing personnel Start-up performance Number of engineering changes Customer feedback Audit deviations Errors and omissions First pass yield
2003	Cheung et al	How well the work is completed in accordance with the design work
2001	Chan et al	Satisfaction level of partnering participants towards the quality of a construction project
2004	Bayliss et al	Counting non-conformance reports and time taken to rectify
2004b,c,d, 2006	Chan et al	7-point Likert scale ranging from very low level to very high

(7) Grand Promenade; (8) Hong Kong Museum of Coastal Defence; and (9) Tradeport Hong Kong Logistics Centre.

Client satisfaction

Client satisfaction is by definition subjective, and as a consequence, is influenced by the individual client's satisfaction (The KPI Working Group, 2000). For this reason, a client satisfaction KPI is developed to address the specific criteria which the client feels are important. In general, the criteria include (1) client satisfaction – product; (2) client satisfaction – service; and (3) client satisfaction – client-specified criteria. It is recommended that the identification of the client-specified criteria and weightings be requested in pre-tender qualifications (The KPI Working Group, 2000). Parkhe (1996) stated that some researchers used subjective measures, such as overall performance assessment by parent companies or JV management to assess JV performance to overcome the limitations of objective measures (they fail to assess qualitative aspects of JV performance). In addition, regular monitoring ought to be conducted in open manner between the client and other participating organizations. This will ensure that the criteria and weightings attached to them are not only relevant and understandable but also the resultant scores are understood, accepted and ultimately acted upon. Table 7.2 shows that 3 out of 17 Hong Kong demonstration projects using an RC approach adopted client satisfaction as a KPI to measure their project performance. These projects are (1) Po Lam Road Phase 1; (2) Chater

Table 7.7 Measures of client satisfaction (adapted from The KPI Working Group, 2000)

Year	Author(s)	Measurement	Definition
2000	The KPI Working Group	Client satisfaction: standard criteria of product	How satisfied the client is with the finished product using the score against the 1 to 10 scale (10 = totally satisfied, 5/6 = neither satisfied nor dissatisfied, 1 = totally dissatisfied)
2000	The KPI Working Group	Client satisfaction: standard criteria of service	How satisfied the client is with the service of the advisor, suppliers and contractors using the score against the 1 to 10 scale (10 = totally satisfied, 5/6 = neither satisfied nor dissatisfied, 1 = totally dissatisfied)
2000	The KPI Working Group	Client satisfaction: client-specified criteria	How satisfied the client is with certain client-specified criteria using the score, against 1 to 10 scale (10 = totally satisfied, 5/6 = neither satisfied nor dissatisfied, 1 = totally dissatisfied), weighted together to determine their level of importance

House; and (3) Lok Ma Chau Viaduct. Table 7.7 indicates the measures of client satisfaction.

Customer satisfaction

Zhao (2002) viewed that customer satisfaction is a KPI for measuring inter-organizational partnerships and he used customer satisfaction rate to measure the performance level of a project. Cheng and Li (2004b) considered that overall satisfaction of stakeholders, including end-users, is one of the three general measures for the success of partnering. They emphasized that the criteria of partnering success are different from those of the project success (always measured in terms of objective project performance in terms of time, cost and subjective project performance in terms of quality) in spite of their possible correlation. The success of RC refers to the perceptive effectiveness of RC by involved parties. This means that the RC arrangement is said to be successful (i.e. achieved effectiveness) if the parties perceive that RC assists to obtain positive outcomes.

Job satisfaction

Job satisfaction refers to the level of individual job satisfaction and career development opportunities. It was used as a KPI in 5 of the 17 Hong Kong demonstration projects with partnering approach. The projects are (1) Three Pacific Place; (2) Po Lam Road Phase 1; (3) Tsim Sha Tsui Station Modification Works, (4) Chater House; and (5) Grand Promenade.

Innovation and improvement

Innovation and improvement is used by Zhao (2002) as a KPI for inter-organizational partnerships. He measured it by counting the number of new initiatives for improvement introduced. In fact, continuous improvement through innovation is a key element for RC (Construction Industry Institute, 1991; Bennett and Jayes, 1998; Walker et al., 2000b). Table 7.2 indicates that innovation is adopted as a KPI in 8 out of 17 Hong Kong demonstration projects using partnering approach. The projects are (1) The Orchards; (2) Three Pacific Place; (3) Cambridge House; (4) Po Lam Road Phase 1; (5) Tsim Sha Tsui Station Modification Works; (6) Stonecutters Bridge; (7) Tradeport Hong Kong Logistics Centre; and (8) East Hall Extension of Passenger Terminal Building of the Hong Kong International Airport.

Relationship-oriented objective measures

Figure 7.5 shows that there are four relationship-oriented objective measures, including: (1) litigation occurrence and magnitude; (2) dispute occurrence and magnitude; (3) claim occurrence and magnitude; and (4) introduction of facilitated workshops.

Litigation occurrence and magnitude

Crane et al. (1999) viewed that litigation is an important relationship-oriented objective measures for partnering project. In fact, litigation is often related to outstanding claims and number of conflicts elevated to each level.

Dispute occurrence and magnitude

Chan et al. (2001) used dispute as a KPI to compare project performance between partnered and non-partnered projects. The result showed that 86.7 per cent of partnered projects had less or equal number of disputes than an average project.

Claim occurrence and magnitude

Claim is adopted by many authors as a KPI for partnering projects (Chan et al., 2001, 2004b, c, d, 2006, Cheung et al., 2003b, Bayliss, 2004). Chan et al. (2001) conducted a study and discovered that 86.8 per cent of the partnering projects had less or equal number of claims than an average project. Bayliss et al. (2004) measured claims by calculating how much time the claims are needed to be settled.

Introduction of facilitated workshops

A demonstration project with partnering approach, Tseung Kwan O Technology Park, indicated that facilitated workshop is a KPI for partnering projects (Table 7.2). As a matter of fact, facilitated workshops are key elements for partnering projects although it is emphasized with less importance (Yeung et al., 2007b). Green (1999) opined that partnering workshops need to be continuous and not one-off at the project start. Walker et al. (2000b, 2002) pointed out that the interviewing process to derive a shortlist of potential alliance members requires sophistication and judgement of a client as do the facilitated workshops. This means that alliance workshops are useful tools to assist to select capable construction alliance partners.

Relationship-oriented subjective measures

Figure 7.5 shows that there are seven relationship-oriented subjective measures, including (1) trust and respect; (2) effective communications; (3) harmonious working relationships; (4) long-term business relationship; (5) top management commitment; (6) employees' attitude; and (7) reduction of paperwork. Like the result-oriented subjective measures, the measurements of the relationship-oriented subjective measures are measured subjectively using a seven-point Likert scale ranging from 1 = extremely low level, 2 = low level, 3 = moderately low level, 4 = neutral, 5 = moderately high level, 6 = high level, to 7 = extremely high level.

Trust and respect

Trust is one of the most important relationship-oriented subjective measures for partnering projects (Crane et al., 1999; Zhao, 2002; Cheung et al., 2003b). Crane et al. (1999) divided trust into internal trust and external trust. Zhao (2002) measured trust by counting the frequency of meeting one's expectations about another party's behaviour and/or having confidence in another party. Wong and Cheung (2004) did a comprehensive study on trust in construction partnering. They identified 14 trust attributes in affecting partners' trust level by a seven-point Likert scale (from 1 = not important to 7 = very important). These trust attributes include (1) reputation; (2) contract and agreements (satisfactory terms); (3) openness and integrity of communication; (4) effective and sufficient information flow; (5) alignment of effort and rewards; (6) adoption of alternative dispute resolution (ADR) techniques; (7) financial stability; (8) frequency and effectiveness of communication; (9) competence of work; (10) sense of unity; (11) problem solving; (12) respect and appreciation of the system; (13) long-term relationships; and (14) compatibility. Table 3.2 shows that 5 out of 17 Hong Kong demonstration projects with partnering approach use trust as a KPI to measure the performance of their partnering projects. These projects are (1) The Orchards; (2) Cambridge House; (3) Tsim Sha Tsui Station Modification Works; (4) Grand Promenade; and (5) Hong Kong Museum of Coastal Defence.

Effective communications

Effective communications is quite often adopted as a subjective measure for partnering projects (Crane et al., 1999; Zhao, 2002; Cheung et al., 2003b; Bayliss, 2004). Crane et al. (1999) divided communication into two types, including internal communication and external communication, to be relationship measures for partnering projects. Zhao (2002) measured communication by counting frequency and type, and calculating the number of information or data exchanges between partners. Bayliss (2004) measured communication by ranking the level of correspondence to and from contractors. Figure 3.3 shows that eleven demonstration projects using partnering approach adopt this KPI to measure their project performance: (1) The Orchards; (2) Three Pacific Place; (3) Cambridge House; (4) Po Lam Road Phase 1; (5) Tsim Sha Tsui Station Modification Works; (6) Choi

Yuen Phase 2; (7) Grand Promenade; (8) Hong Kong Museum of Coastal Defence; (9) Tradeport Hong Kong Logistics Centre; (10) Tseung Kwan O Technology Park; and (11) One Peking Road.

Harmonious working relationships

Chan et al. (2001) used satisfaction level of working relationship as a KPI and they found that 78.2 per cent of the partnering participants strongly agreed that they were happy with the working relationship. Cheng and Li (2004b) stated that subjective measures are based on the notion that partnering is used to improve working relationships that help to achieve predetermined common goals for fulfilling the overall satisfaction of stakeholders. Therefore, an improved or harmonious working relationship is one of the general measures of the success of partnering. Table 3.2 indicates that five Hong Kong demonstration projects using partnering approach use this KPI to measure their project performance: (1) Three Pacific Place; (2) Chater House; (3) Stonecutters Bridge; (4) Grand Promenade; and (5) Design and Build of Improvement to Castle Peak Road.

Long-term business relationship

Long-term business relationship is used as a KPI in five of the Hong Kong demonstration projects with partnering approach (Table 3.2). These projects included: (1) The Orchards; (2) Three Pacific Place; (3) Cambridge House; (4) Tsim Sha Tsui Station Modification Works; and (5) Choi Yuen Phase 2.

Top management commitment

Although top management commitment is often viewed as a critical success factor for RC projects (Harback et al., 1994; Slater, 1998; Cheng et al., 2000; Chan et al., 2004b), it can also be viewed as a KPI because it is both means and ends for RC projects, depending on which perspective one considers. Table 3.2 indicates that a demonstration project with partnering approach, Lok Ma Chau Viaduct, uses top management commitment as a KPI to measure its project performance.

Employees' attitude

Employees' attitude refers to their attitude towards the implementation of RC approach of a project. Zhao (2002) applied it as a KPI for inter-organizational partnerships which were measured by employee turnover rate.

Reduction of paperwork

A demonstration project with partnering approach, Chater House, uses reduction of paperwork as a KPI for measuring its partnering performance (Table 3.2). In fact, it is generally agreed that efficiency of communication through partnering was enhanced (Chan et al., 2004b, c, d, 2006), and practitioners were able to get faster responses by having more informal communication and the potential problems were reduced immediately by open communication. Therefore, level of paperwork reduction can reflect level of effective communication.

Chapter summary

This chapter has conducted a comprehensive literature review on performance measures for construction RC projects and a preliminary conceptual framework for identifying KPIs for construction RC projects has been developed. To verify the validity of this conceptual framework, an in-depth study of 17 demonstration projects using partnering approaches (out of 24 projects in total) derived from the Hong Kong Demonstration Committee was conducted. A consolidated conceptual framework for identifying KPIs for construction RC projects has thus been established. A total of 25 performance measures have been sought and discussed within this conceptual framework and they are classified into four major categories, including, (a) result-oriented objective measures; (b) result-oriented subjective measures; (c) relationship-oriented objective measures; and (d) relationship-oriented subjective measures. This KPI conceptual framework for RC projects can assist in setting a benchmark for measuring the performance of RC projects. Previous research on performance measures for different forms of RC proved that the above-mentioned performance measures for construction RC projects are sound and valid. As a result, construction senior executives and project managers can apply them to measure and assess the performance of their RC projects.

Part II
Relational contracting practices

Chapter 8

Relational contracting approach and process

Introduction

To implement a construction project successfully, a key is to develop a good delivery strategy. In fact, whether it is better to adopt an RC approach depends on certain criteria. Main Roads Project Delivery System (2005) stated there are three main decisions a client must make to create a suitable delivery strategy. The items encompass: (1) physical packaging of the project; (2) selection of suitable project delivery method; and (3) risk framework. Key elements of physical packaging of a project include: (1) economy of scale; (2) bulking up like works; (3) sequencing of the works; (4) specialization; (5) supply of materials; (6) industry liaison and large complicated projects. To select a suitable contract type, it is useful for a client to develop a project assessment tool. This project assessment tool should comprise: (1) scope of the project; (2) timing; (3) risk; (4) constructability; (5) sensitivity; (6) capacity and capability; (7) budget; and (8) location (Main Roads Project Delivery System, 2005).

Physical packaging of a project

Main Roads Project Delivery System (2005) stated that there are various issues a client should consider when deciding how to package and deliver a project. The issues are:

1   How the delivery method of a project can assist in the government's broader priorities;
2   The number and size of packages;
3   Vertical, horizontal or product separation of the packages;
4   The most suitable contract type or process to adopt;
5   The most suitable tendering or selection process to adopt;
6   The most suitable remuneration method;
7   Government policy with regard to support for local communities.

The selection of the most suitable delivery method should be considered in the conceptual phase of the project cycle so as to maximize profits. In some highly complicated projects, it may be appropriate to use group problem-solving workshops to determine the packaging and delivery method to be adopted. Complexity and uncertainty in the conceptual phase do not necessarily mean complexity and uncertainty in the finished product. However, clients need to be able to explain critical factors such as time, cost, quality, environmental and cultural restraints and their effect on the community and local businesses (Main Roads Project Delivery System, 2005). It should be noted that risk will increase when complexity and the degree of uncertainty increase. In addition, it may be suitable to develop an alliance contract with suppliers if a project is highly complex with a high degree of uncertainty. Figure 8.1 shows that when complexity and the degree of unknowns increase, the need for collaboration and good relationships increases simultaneously.

Selection of suitable project delivery method

The development of a project assessment tool which is based on common sense and experience, can help to select a suitable delivery method. Table 8.1 indicates items that ought to be considered when determining the most appropriate method of delivery. For large projects, the project assessment tool may be used as quick guide. Nevertheless, it is better to follow up with a value management workshop (Main Roads Project Delivery System, 2005). Additional industry representatives, such as product suppliers and quarries may benefit from some projects. Table 8.2 examines a proposed project in relation to various table headings. Some questions that are not directly applicable should be deleted but some other appropriate questions should be added. Then, a weighting factor should be assigned for each category with the sum of all factors equal to 100%. In more detail, the assignment of weights may be through professional judgement or a system known as pair-wise comparison (Main Roads Project Delivery System, 2005).

Risk framework

Before selecting project packaging and works delivery methods, it is essential to document the basic characteristics of a project and make an assessment of the project risk profile. The characteristics of project in general include: (1) estimated cost; (2) project classification and location; (3) site conditions; (4) availability of preliminary design; (5) timing considerations; and (6) specific works elements. The project risk profiles that are common to all projects include: (1) over-run budget; (2) late completion; (3) not satisfying all regulatory bodies and stakeholders; and (4) not conforming to standards, drawings and specifications as written in a contract.

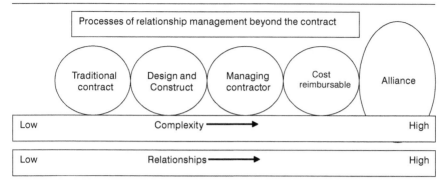

Figure 8.1 Relationship management (reproduced from Main Roads Project Delivery System, 2005) (with permission for both print and online use from the Department of Main Roads, Brisbane, Australia)

In fact, risk should be managed by parties who are best able to handle them (Main Roads Project Delivery System, 2005), for example:

1   A client is best able to manage, and should retain responsibility for, risks associated with: the minimum design and construction standard of the infrastructure; community and governmental acceptance; environmental impact remediation; right of way matters.
2   The contractor is best able to manage risk associated with: construction and maintenance, encompassing conformance with construction standards; safety of workers during construction.
3   The designer is best able to manage risks associated with providing a design that will: achieve the client's purpose and follow the client's design and construction standards.

Approaches and processes of implementing different types of relational contracting

As mentioned in previous chapters, there are four major types of RC, including (1) partnering; (2) alliancing; (3) PPP; and (4) JV. In fact, the approaches and processes of implementing them are quite different and the following sub-sections illustrate them in detail.

Partnering approach

Partnering is a form of relationship highly valued by many industrial participants and presents an ideal opportunity to enhance harmonious working relationships (Chan et al, 2004b). In fact, there are both structured and unstructured approaches when implementing partnering. In the structured approach, there are a number of partnering workshops encompassing: the initial workshop; interim workshops; and a final workshop. In contrast,

Table 8.1 Project assessment tool (reproduced from Main Roads Project Delivery System, 2005) (with permission for both print and online use from Department of Main Roads, Brisbane, Australia)

Category	Questions	Weighting
Scope	• Is the scope able to be well defined? • Can the scope be fully detailed? • Is the scope likely to change during detailed design/construction? • Does the project have additional objectives beyond time, cost, quality and safety? • Does the scope include operations/maintenance?	
Time	• Are there seasonal, externally imposed constraints on delivery? • Is early completion of value to the client? • Is it essential and/or possible to fast track? • Is timetable able to be negotiated? • Will late project delivery produce bad consequences?	
Budget	• Is there a requirement to meet a guaranteed maximum price?	
Risk	• Is risk to be clearly defined? • To what extent can risks be mitigated? • Is the risk profile best shared between owners and contractors?	
Constructability	• Is the required technology proven/new? • Is the technology and materials widely practiced/available from different sources? • Are construction processes simple or routine? • Is construction input to design imperative to success? • Is design input to construction imperative to success? • Will packaging improve constructability?	
Sensitivity	• Is the project environmentally and culturally sensitive? • Are local stakeholders/business sensitive? • Are sensitivities able to be clearly defined? • To what extent can sensitivities be mitigated? • Is the sensitivity profile best shared between owners and contractors?	
Capacity and capability	• Are adequate contractors' resources available in the marketplace? • Will the project need special contractor capabilities? • Are sufficient owner resources available? • Is the owner willing/capable to be part of an integrated team?	
Total		100%

Table 8.2 Example only – Scope category element (reproduced from Main Roads Project Delivery System, 2005) (with permission for both print and online use from Department of Main Roads, Brisbane, Australia)

Question	Weighting	Description of low rating	1	2	3	4	5	6	7	Description of high rating
1. Is the scope able to be well defined?		Well defined		2						Poorly defined
2. Can the scope be fully detailed?		Full detail		2						Nil detail
3. Is the scope likely to change during detailed design or construction?	15%	Low likelihood	1							High likelihood
4. Does the project have additional objectives beyond time, cost, quality and safety?		One objective				3		8		Many objectives
5. Does the scope include operations or maintenance?		No	1							Yes
			2	4		3		8		Total = 17

Add drop down sub-total
Scope

→

the unstructured approach impedes the partnering spirit in a construction project. As a matter of fact, the process of partnering could become the culture on all projects regardless of the delivery method because partnering is not a form of contract. Instead it is a process beyond the contract to align the common goals and objectives of the parties to the contract and to facilitate effective communication, integrated teamwork and joint problem solving. It should be noted that partnering can be used together with any form of contract (Main Roads Project Delivery System, 2005); the partnering charter will not have any legal binding. It is often viewed that partnering is simply a way of doing business in which two or more organizations make long-term commitments to achieve mutual goals (Construction Industry Institute, 1991). Main Roads Project Delivery System (2005) stated that partnering is a relationship in which:

1   All participants are expected and encouraged to have trust and open communications between them;
2   All parties can resolve issues and problems quickly and at the lowest possible level;
3   Parties try to develop mutually agreed solutions that meet the needs of all parties;
4   Common goals are identified by all parties working for the project;
5   Partners seek input from each other to find better solutions to the current problems and issues.

Partnering process

Partnering is a relationship-based procurement method aimed at avoiding contractual disputes, reducing litigation, improving trust between contractual partners and building inter-organizational teams (Chan et al., 2006). If partnering is implemented properly, it can achieve time and cost savings, improve safety and value engineering efforts, reduce litigation and result in significant improvement in achieving contractual goals.

The following partnering process is a general outline that should be implemented (Main Roads Project Delivery System, 2005).

1   Select a suitable contract on which to adopt partnering:
     Constant communication is required for complicated, difficult and high-risk contracts;
     Long-term contracts will reap the greatest benefit;
     Depth of the partnering effort varies with the complexity of a contract;
     Can be used with maintenance contracts as well as construction ones.
2   Obtain senior management commitment:
     Workers pay attention to how senior management commits;
     If senior management is not committed, it will be hard to obtain resources or to change the day-to-day way of doing business;

Senior management ought to sign the partnering agreement to signify their commitment to the project and to the whole partnering process.

3   Obtain essential resources to partner. Err on the high side: budget for resources to include the cost for workshops, and the time commitment to train personnel, have them attend workshops and for clear communication between parties.

4   Select partnering champions from each organization: one at senior level and one at working level. These people are responsible for promoting the partnering philosophy throughout the organization and help the key participants change their actions to support the partnering culture.

5   Encompass a partnering clause in the contract. Let the contractor know that you are offering partnering as a way to improve working relationships between parties.

6   Reinforce your commitment to partnering at all meetings, and in correspondence and verbal communications between the client and contractor.

7   Consent to partner:
All processes have to be mutually agreed by all parties;
It is better for initial contacts to be at executive level;
It is good to establish an executive-level relationship;
Mutual high-level commitment should be ensured.

8   Recruit a facilitator:
Add to the partnering charter that the facilitator is a requirement;
Internal personnel should be familiar with the process.

9   Prepare for an initial partnering workshop:
Get support from top management so that failure is not intimated at the workshop;
Ensure all attendees understand the basic principles of partnering;
Identify all participants in the programme;
Clarify the roles, responsibilities and authorities of all key players.

10   Prepare for interim partnering workshops:
Review the partnering performance monitoring matrix;
Evaluate and identify perceived benefits of partnering;
Evaluate and identify significant difficulties in implementing partnering;
Evaluate critical success factors for partnering implementation;
Suggest remedial actions to overcome identified obstacles;
Develop an issue resolution mechanism and re-nominate partnering champion.

11   Generate a meaningful evaluation process:
Measure success of the parties;
Reinforce the partnering attitude;
Use evaluation sheets;
Measure regularly, i.e. every six months.

12   Celebrate successes. Reward or recognize parties who took the initiative to partner, whether successful or unsuccessful.
13   Communicate regularly with your partner:
Regular meetings to discuss progress, problems and solutions are recommended;
Face-to-face contact is recommended.
14   Other considerations:
Prepare to address newcomers to the programme. Do not assume that they will fall right into the partnering pattern;
Measure the success rate, and adjust processes when necessary to deal with problem areas;
Establish a means for regular social interaction at the workplace such as morning teas, lunches, birthday parties, etc;
Follow management philosophy established at the workshop. Partnering in this manner has proven to produce dramatic improvement in contract success.

Main Roads Project Delivery System (2005) viewed that a typical agenda for partnering should include the following items: (1) purpose; (2) objectives; (3) workshop content; (4) workshop participants; (5) duration; (6) relationship management meetings; and (7) meeting content. These meetings should normally last about two hours and are held before the monthly progress meetings.

1   Purpose. To kick off a collaborative approach to the delivery of a project, and to develop a partnering charter and a systematic approach to the management of relationships within the contract.
2   Objectives:
Define common goals and objectives for the delivery of the project;
Begin training team members;
Organize continuous relationship management meetings.
3   Workshop content:
Introduction and welcome;
Ice breaking;
Each member introduces another;
Roles in the project;
Expectations;
Grounding presentation – the context;
Team development theory;
Personality analysis;
Develop the mission statement;
Brainstorm keywords individually;
Small groups work to put keywords together;
Group as a whole works on statement;

Determine the mission objectives;
Split into groups by party to the contract to determine
critical success factors for launching partnering;
Bring together groups and view objectives;
Find commonality and combine objectives;
Agree on partnering charter;
Continuous assessment of objectives;
Overview of performance monitoring of the objectives;
Determine the administrator of the partnering process;
Determine the chair;
Determine the attendees;
Develop a typical agenda;
Risk assessment;
Brief on features of project;
Brainstorm in mixed groups;
What is working well;
What is not working well;
What strategies can we put in place to fix these;
Group report to whole group;
Opportunity assessment;
Brainstorm opportunities;
Develop strategies to implement;
Develop issues resolution matrix;
Discussion of types of issues;
Big issues dealt with separately;
Explain the ground rules;
Overview of skills workshop;
Develop action plan;
Sign partnering charter.

4   Workshop participants:
Client's team;
Consultant's team;
Contractor's team including major subcontractors;
Superintendent's team.

5   Duration should be 1.5 days with a dinner session

6   Relationship management meetings:
Purpose: to monitor, identify and resolve issues regarding the
relationships on the project;
Objective: to monitor the common objectives developed and agreed
for the project; to raise issues relating to relationships on the project;
to resolve issues relating to relationships on the project.
Meeting participants as agreed at partnering foundation workshop,
typically contractors, project manager and site supervisors;
representatives from major subcontractors with work currently

in progress; superintendent, superintendent's representative and inspectors; client's representative; and designer.

7   Meeting content

Welcome and introduction: review actions arising from last meeting;

Current month's scores relating to key objectives and brief discussions regarding each score;

Trends in the scores regarding objectives: monthly trend, comparison with previous month, identification of areas of improvement; identification of areas which are not meeting our expectations.

General comments: what is going well; what is not going well; develop strategies.

Develop action plan.

Next meeting

Extended partnering approach

Main Roads Project Delivery System (2005) introduced extended partnering, a business model that evolved from the Pacific Motorway project. It is an extension of the process introduced by the Relationship Management Unit from that project. In fact, extended partnering is a formal process adopted to allow greater team participation and communication beyond the contractual process. It is used to develop a cooperative working approach between all parties to the contract to achieve the best project outcomes. It does this by achieving some initial impetus towards achieving positive working relationships and shared goals at the start of the project, providing appropriate problem solving and team skills to the parties to achieve and maintain these and by providing an agreed structure, which assists in maintaining and improving these relationships. Figure 8.2 shows various elements of the extended partnering process. This procurement approach can be applied to different traditional types of contracts, design and construct, design construct and maintain, managing contractor, and cost reimbursable performance incentive. It would normally be used more frequently for complex projects.

Extended partnering process

It is of interest to note that the partnering process and agreements formed in the extended partnering process do not replace the requirements of the contract because there is still a requirement to adhere to contractual notices, timelines and directions (Main Roads Project Delivery System, 2005). However, information flow is faster and more comprehensive than that provided for in the contract, issue resolution durations are far shorter and there is more focus on joint objectives rather those of the individual parties (Main Roads Project Delivery System, 2005).

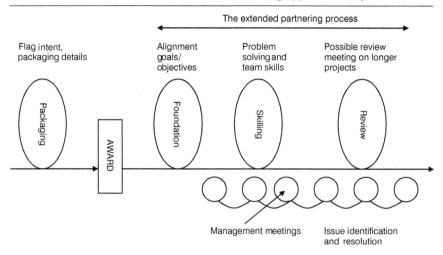

Figure 8.2 A typical 'extended' partnering process (reproduced from Main Roads Project Delivery System, 2005) (with permission for both print and online use from the Department of Main Roads, Brisbane, Australia)

In general, a typical extended partnering process as implemented by the Department of Main Roads is as follows (Main Roads Project Delivery System, 2005):

1   Provision of an intent to proceed with the relationship management process either in the invitations for tender or in the contract documents.

2   A foundation workshop typically lasting for 1.5 days early after the award of the contract to align various contracting parties towards common goals; to develop performance measures for these goals; to develop an issue resolution process; to start communications and relationships on a positive footing; and to develop plans and process for further managing this.

3   A series of relationship management meetings of (approximately two hours long); held at approximately monthly intervals to monitor the common goals, promote communication, raise and discuss issues, and develop plans to resolve them.

4   A skills workshop is held about four to six weeks after the foundation workshop to provide team-building, communication, motivation, problem-solving and decision-making skills to all project parties. The duration of this workshop is approximately 1.5 days.

The boundary rider

Main Roads Project Delivery System (2005) stated that it is advisable to nominate a boundary rider to the process. The boundary rider ought to be independent of the 'team' formed to deliver the project. The role of the boundary rider is to observe and comment on team dynamics from time to time and to be on call as facilitator for more difficult problems as they arise during the project delivery. If team dynamics are good and significant problems do not arise in a project, there is no need to call upon the boundary rider. If the duration of a project is relatively short, it is usually not practical to provide a foundation and skilling workshop within the timeframe of the contract. In such cases, an expanded foundation workshop may be provided. Monthly management meetings are held as for the extended partnering process, and certain elements of the skilling workshop are thus introduced at the foundation workshop.

Alliancing approach

Project alliancing aims to create mutually beneficial relationships between all parties involved to produce outstanding project outcomes (Queensland Department of Main Roads, 2005). Unlike traditional forms of contract where risk is allocated to various parties, the alliance participants take collective ownership of all risks associated with the delivery of the project, with equitable gain-sharing/pain-sharing (in fixed pre-agreed ratios), depending on how the outcomes compare with pre-agreed targets (Walker et al., 2000; Queensland Department of Main Roads, 2005). The risk-reward arrangements are designed to achieve excellent outcomes for all parties. This underlying commercial alignment is consistent with an alliance philosophy that focuses all parties on achieving common objectives to attain a win-win situation. In fact, before adopting an alliancing approach, a client should decide whether this method gives value for money. It should be noted that the consequences of using this procurement method inappropriately are potentially more significant than for other forms of procurement. Also, the procurement decision ought to be based on an analysis of how the risks and opportunities associated with a project can be managed for each available procurement method (Department of Treasury and Finance, 2006). Project alliancing is typically suited to projects with the following characteristics:

1   Numerous complicated and/or unpredictable risks with complex interfaces;
2   Complicated stakeholder issues;
3   Complex external opportunities and threats that can only effectively be managed collectively;
4   Very tight timeframes;

5   Output specifications which cannot be clearly defined upfront, and/ or a high likelihood of scope changes during design and construction;
6   A need for owner involvement or significant value adding during delivery.

In contrast, alliancing is in general not suitable to projects where:

1   Risks can be clearly defined and allocated, without the need for the client's involvement;
2   The project offers significant whole-of-life efficiencies and opportunities, and these would not be available if an alliance is used;
3   The project is of a scale where any benefits that could be achieved from using project alliancing would be offset by increased procurement costs associated with that method;
4   The case for alliancing is marginally benefited when compared with other procurement methods.

A recommended process in confirming the appropriateness of alliancing is illustrated in Figure 8.3. There are three important steps to confirm the appropriateness of alliancing as suggested by Department of Treasury and Finance (2006), which included: primary tests; analysis of procurement methods against project objectives; and supplementary comparative assessments.

Alliance principles

After deciding to adopt alliance to procure a construction project, it is important to note the following vital alliance principles (Queensland Department of Main Roads, 2005):

- A primary emphasis on business outcomes whereby all parties either win or lose;
- Collective responsibility for performance with an equitable risk and reward sharing;
- A peer relationship where all participants have an equal say;
- All decisions must be best for the project;
- Clear responsibilities within a no-blame working culture;
- Full access to the resources, skills and expertise of all parties;
- All transactions are fully open book;
- Encouragement of innovative thinking with a commitment to achieve outstanding outcomes;
- Open and honest communication, i.e. no hidden agendas;
- Visible and unconditional support at all levels of each participant.

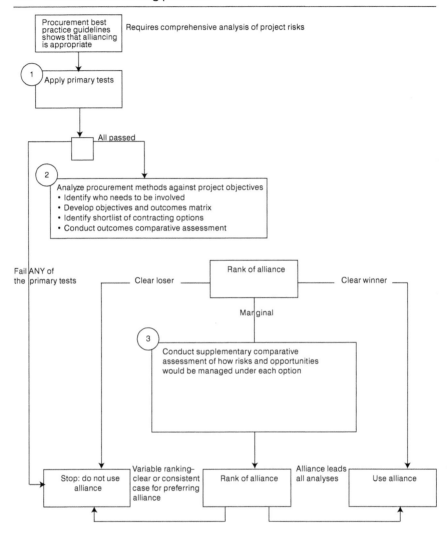

Figure 8.3 The steps in the decision-making process (adapted from Department of Treasury and Finance, 2006)

Alliancing process

After making a decision to use alliance, the most critical step for a client is to set up the alliance framework properly and select the right participants to join the alliance and deliver the project. There are two distinct stages for implementing alliance process: (1) request for proposals (RFP) development: develop the alliance model and invite submissions from proponents; (2)

evaluation and selection: selecting the non-owner participants (NOPs). The key activities in each stage are summarized in Table 8.3.

In most cases, the client uses some form of competitive process to select the NOPs for the alliance. Nevertheless, there are some situations where this may not be the case and the client may need to engage the NOPs on a 'sole-source' basis.

Overview of request for proposals development stage

After deciding the form of the alliance and the evaluation and selection process, the client issues an RFP giving details of the project, the alliance, the

Table 8.3 Stages in establishing the alliance framework (adapted from Department of Treasury and Finance, 2006)

	Establish the alliance – Develop alliance model; select NOPs	
	Develop model and invite proposals	Select NOPs
	Key activities (by proponents) • Learn about project/owner's needs • Identify best available resources • Start building the project team	Key activities (by owner) • Read/evaluate written submissions • Conduct interviews and workshops • Select preferred proponent • Initiate/conduct establishment audits • Conduct commercial alignment process • Further development of owner team • Continue work on project issues
Decision to use alliance	Key activities (by client) • Engage specialist advisers (alliance/commercial, legal, probity) • Decide on form of alliance framework • Develop selection process Secure sign-off on process • Develop/commit to schedule Set up and train selection panel • Develop criteria and scoring guidelines • Develop all EOI/RFP documentation • Recruit/select owner's team nominees • Educate/prepare owner's team • Notify industry of intentions • Start selection of FA-E, IE	Key activities (by proponents) • Prepare written submission • Develop/prepare project team • Participate in interviews and workshops • Assist with establishment audits • Participate in commercial alignment process

evaluation and selection process, and inviting suitably qualified proponents to submit proposals (Department of Treasury and Finance, 2006). There are seven essential steps leading to an RFP:

Step 1: develop/agree overall approach;
Step 2: develop and commit to establishment schedule;
Step 3: establish selection panel and evaluation procedures;
Step 4: develop the RFP documents;
Step 5: develop the owner's team;
Step 6: early notice to industry; and
Step 7: engage financial auditor, an independent estimator.

Figure 8.4 summarizes the key activities leading up to the issue of the RFP.

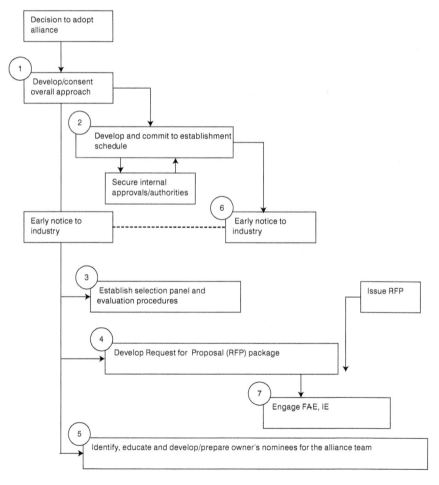

Figure 8.4 Steps leading to issuing a request for proposals (adapted from Department of Treasury and Finance, 2006)

The duration of the RFP stage varies, depending on the circumstances. There may be circumstances where it is essential or suitable to start the process with a request for expressions of interest (EOI) before the RFP; for example, where the client is uncertain about the level of capability/interest in the marketplace; or if there is a concern that there will be too many responses to an RFP and it would be less wasteful for parties to filter the number of proponents through an EOI process.

Unless there is a compelling reason for an EOI stage, it is recommended that the client goes straight to an RFP.

Overview of the selection process

The selection process should be designed and implemented to develop the right psychological foundation for the eventual alliance. The right process can help to generate momentum for the launch of the alliance and minimize the cost of the process to the owner and to industry (Department of Treasury and Finance, 2006). Figure 8.5 shows a typical recommended process for selecting the NOPs.

The above-mentioned process is designed to ensure that the client meets the team directly involved in leading and delivering the project. It is suggested that the client uses a specialist alliance facilitator to ensure that the client's team and the selection panel understand thoroughly the subtleties of the process and have the insights and skills to implement the process effectively. Table 8.4 shows a typical timetable for the selection process.

It is becoming more common for a client to request and agree that some project development works get underway as soon as the preferred participants have been selected. In this case, the 'early start work' proceeds in parallel with the establishment audits and commercial alignment process, before the execution of the alliance agreement (Department of Treasury and Finance, 2006). The scope of this early start work is often limited to just a few of the key players from the owner's team to better understand the project and start establishing some of the systems and processes for the alliance project office. Those involved must ensure that all approvals are satisfied and in place before agreeing to any early start work for which the client may be obliged to pay (Department of Treasury and Finance, 2006).

Other considerations

There are some other considerations that need to pay attention, including using an individual or consortium approach; use of specialist advisers; and selection on sole-source basis (Department of Treasury and Finance, 2006). The term 'sole-source basis' means that there is only one team of NOPs applied for the alliance.

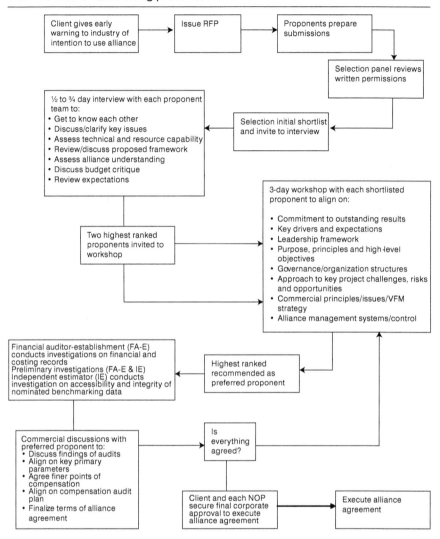

Figure 8.5 Recommended process for selecting alliance NOPs (adapted from Department of Treasury and Finance, 2006)

In fact, the client needs to decide whether to undertake a separate selection process for each of the key participants (e.g. designer, constructor and so on) or to allow industry to form its own allegiances and make submissions as a consortia incorporating the required range of capabilities. The most serious drawback for the client in allowing industry to form its own consortia is that the client's choice is limited to the consortia that result. The client does not have the option to mix and match. It has to choose the best of the consortia

Table 8.4 Typical timetable for selecting alliance NOPs (adapted from Department of Treasury and Finance, 2006)

	Typical duration
Issue RFP	
Proponents prepare teams/written responses	4-5 weeks
Time allowed for proponents to register intention	1 week
Proponent briefing	2-3 hours
Receive written submissions	
Evaluation of submissions/reference checks	2 weeks
Conduct interviews with shortlisted proponents	1 week
Notify/invitations to workshops	
FA-E (Financial Auditor/Estimator) briefing to proponent financial officers	½ day
First 2-day selection workshop	2 days
Second 2-day selection workshop	2 days
Selection/advise preferred proponent	
FA-E and IE (Independent Estimator) kick-off meetings	½ day
Start establishment audits/IE preliminary review	5-8 days
Complete and report on establishment audits	2 weeks
Commercial discussions/alignment	2 weeks
Alliance agreement ready to sign	
Final approval from client to proceed	varies
Early start work	

offered, in spite of concerns about any of its members. However, while the alternative approach of conducting separate selection processes allows the client to seek the best individual companies, it has many disadvantages. The individual process takes longer and although the client might get the best individual companies, it may not get the best team because the two selected teams may not be a good cultural fit. The client does not know this until after the selection is made. Under the consortia approach, the companies will have already bonded into an effective team by the time they engage with the client's team. Therefore, in most cases, the client chooses to adopt the consortia approach.

In addition to a probity adviser/probity auditor, the client will need to engage the assistance of certain specialist advisers, who may encompass some or all of the followings: (1) alliance adviser/facilitator(s); (2) alliance legal adviser; (3) financial auditor(s); and (4) independent estimator(s). In addition, the client may need to engage the services of an insurance specialist.

On the other hand, the client in most situations needs to adopt the competitive process to select the NOPs for the alliance. There are some other situations, however, where this may not be the case. For example, only one company/consortium may be capable of delivering the project and the owner may need to select on a sole-source basis. In such cases, the client still needs to undertake some of the selection process steps to ensure that the alliance is established on a sound commercial and psychological foundation (Department of Treasury and Finance, 2006).

Evaluation and selection

Once the written submissions have been received from the proponents, selection of the preferred participants involves the following key steps (Department of Treasury and Finance, 2006):

1   review/score written submissions: select initial shortlist;
2   interviews: select final shortlist;
3   selection workshops: decide preferred proponent.

Commercial alignment

When the preferred participant has been selected, top management personnel from the client and the preferred participant need to conduct a series of meetings/workshops to reach alignment on the commercial framework for the alliance and to agree on the terms of the alliance agreements. The key outcomes from the commercial discussion are (Department of Treasury and Finance, 2006):

- Alliance agreement ready for signing;

Table 8.5 Typical timing for selecting the preferred proponent (adapted from Department of Treasury and Finance, 2006)

	Typical duration
Receive written submissions	
Evaluation of submissions/reference checks	2 weeks
Conduct interviews with short-listed proponents	1 week
Notify/invitations to workshops	
FA-E a briefing to proponent financial officers	½ day
First 2-day selection workshop	2 days
Second 2-day selection workshop	2 days
Selection/advise preferred proponent	

- Agreement on the limb-3 framework and all primary parameters of the pain-gain sharing model, encompassing an outline of the key responsibility (KRA) and KPI framework and measures. Although agreement should be reached on as many aspects of the limb-3 framework as possible, there will be certain aspects that cannot be resolved until during the project development phase. It may be suitable to seek provisional alignment on certain aspects in the commercial discussions, on the understanding that these will be reviewed and finalized during the project development stage;
- Agreed budget for the project development phase;
- Agreement in relation to the specific value for money initiatives outlined in Table 8.6.

It should be noted that there are some important steps for the commercial alignment:

Step 1: issue briefing papers;
Step 2: FA-E briefing to financial officers; (3)
Step 3: auditor/IE kick-off meeting;
Step 4: establishment audits and preliminary investigation; and
Step 5: commercial alignment discussions.

Alliance implementation issues

Once established, it is expected that an alliance will use the best practice business and project management systems and processes through all stages of project delivery. The following aspects should be addressed in alliance implementation (Department of Treasury and Finance, 2006).

- Overview of purpose and outcomes;
- Alliance management plan;

Table 8.6 Commercial framework for client/preferred proponent agreement (adapted from Department of Treasury and Finance, 2006)

Items to consent
Details of different value for money deliverables
The FA-E report encompassing:
1. all details of limb-1 costing
2. limb-2 fee% and methodology for applying them
3. draft compensation audit plan
The proponent's budget critique
Confirmation by the independent estimator that the data from nominated benchmarking projects is accessible, relevant, creditable and available.

- Develop and agree the target cost estimate (TCE) and target out-turn cost (TOC);
- Effective project governance;
- Develop and sustain alliance culture;
- Closure.

Project alliancing can be divided into three phases, including (1) project development phase; (2) implementation phase; and (3) defects correction period (Department of Treasury and Finance, 2006).

During the project development phase, the alliance is focused on getting the team established, defining and/or clarifying the scope of work and agreeing the TOC and other performance targets for the implementation phase. Some key issues need to be addressed, including:

1   Establish the project team, especially: (a) logistics: office, administration, network systems, communication protocols; and (b) people: set-up organization structure and appoint personnel, role responsibilities and accountabilities, team culture;
2   Secure all essential approvals;
3   Define the scope of work: the process could vary considerably depending on the circumstance;
4   Develop an alliance management plan incorporating the different policies and plans that will be used to guide and manage the project;
5   Prepare the TCE and agree the TOC;
6   Develop and agree the KRA/KPI framework, including details of all targets and measurement methods;
7   Procure all necessary insurances for the implementation phase;
8   Conduct workshops and other processes necessary to develop a high-performance alliance culture.

The focus throughout the implementation phase is on meeting or exceeding the agreed project objectives. Key issues to be addressed during implementation include leadership, governance, management controls and reporting, and sustaining peak performing alliance culture.

During the defects correction period, the alliance is responsible for attending to any defects that emerge in this period. This period may also be used to measure different aspects of the performance of the facility under operating conditions. These measures may be linked to the KRA/KPI framework (Department of Treasury and Finance, 2006).

PPP approach

Main Roads Project Delivery System (2005) stated that international experience of the adoption of PPP can provide real benefits in terms of value for money outcomes in construction and continuous delivery/provision of transport infrastructure services. These benefits are not only tied to cost savings and innovations during construction but also efficiencies gained in the maintenance and delivery of transport infrastructure services.

PPP encompasses a broad spectrum of project delivery options and it is often alternatively referred to as private finance initiatives (PFI). PPP is about the delivery of public services through partnerships between governments and the private sector (Asian Development Bank, 2007). The private sector provides ancillary services with the core service provided by government. The government holds end-user risk. It should be noted that PPP does not alter a government's responsibility for policy or the delivery of services to the community. It aims to achieve value for money, allow the community to benefit from innovation obtained from private-sector investment and skills, and provide new infrastructure and services that may not otherwise be available due to governmental budget constraints. As a matter of fact, the government has to thoroughly work out and clearly state its own objectives for the project so as to sufficiently plan and implement a successful engagement process. It is also critical to understand the private sector forces that drive consortia bids (Main Roads Project Delivery System, 2005).

A key to success of PPP is the efficient allocation of risk between the public and private sectors, with the underlying principle being that individual risks should be borne by the party best able to manage them (Asian Development Bank, 2007). Typical risks transferred from the public sector to the private sector consortium might include risks associated with design and construction of infrastructure, availability or performance, long-term maintenance, technology, and financing.

Main Roads Project Delivery System (2005) used a method for relating value for money and risk transfer and concluded that public procurement would benefit if certain risks could be transferred to other parties. An optimum value for money position could be reached, a point beyond which it was hard for the private sector to price or absorb further risk, however, conventional procurement was indicated to be at a less than optimal position with more risk transference desirable. Figure 8.6 shows an indicative representation on how risk is transferred.

The generally accepted procedure for managing a potential PPP involves constructing a risk matrix that identifies: (1) each risk; (2) causes of risk; (3) consequences of risk; (4) potential financial impact; and (5) strategy for mitigation. In fact, PPP projects have certain risks such as design and construction risks, approvals, finance, economic downturns, political decisions, technology changes, operational, market, maintenance and residual values (Main Roads Project Delivery System, 2005). Therefore, PPP should provide governments and private consortia with a strong incentive to focus on whole-of-life service delivery and costs, rather than taking a short-

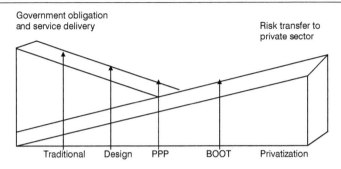

Figure 8.6 Type of contract and risk transfer (reproduced from Main Roads Project Delivery System, 2005) (with permission for both print and online use from Department of Main Roads, Brisbane, Australia)

term view linked to the current government. For this reason, the consortium has to be composed of operators and asset builders and financiers.

PPP process

Before adopting the PPP approach to procure a construction project, a feasibility study ought to be conducted if a client has identified a construction project that is required and is considered to have potential as a PPP. A decision to proceed with the PPP approach would often follow from the feasibility study recommendations which should address the following items (Hong Kong Efficiency Unit, 2003):

1   Provide service specifications, which can be agreed upon by stakeholders and tested in the market;
2   Assess the financial viability of the project;
3   Whether the client has the essential legal powers to let successfully a contract for the provision of the service in question and, if not, what can be done to address the absence of such powers.

After conducting the feasibility study, the client should conduct technical assessments before inviting proposals. Like any project, there is a need for community endorsement of a PPP project proposal. For publicly funded PPP projects, a public works programme (PWP) item should be created. Nevertheless, it is equally vital for a number of reasons that a PWP item ought to be created for privately financed projects, encompassing those that are expected to be finally free standing (Hong Kong Efficiency Unit, 2003):

1   To ensure that funding support is secured for any capital or recurrent implications arising from the project;

2   To enable different departments to proceed with the various technical and other studies that different departments need to conduct prior to the RFP;

3   To demonstrate to potential bidders that government can procure the project in the conventional way if no value for money PPP bids are received or the PPP process does not result in a satisfactory outcome.

Hong Kong Efficiency Unit (2003) stated that there are other important points that need to take note when implementing the PPP process, including that it is not worthwhile to procure PPP projects for small value projects and the duration selected should result in best value for money; it is generally efficient if the contract length is somewhere between a half and the whole of the useful life of the main assets.

JV approach

Geringer (1988) stated that in a JV there are two or more legally individual organizations (the parents), and each shares in the decision-making activities of the jointly owned entity. Tomlinson (1970) defined JV as an arrangement where there is commitment of funds, facilities and services by at least two legally separated interests to an enterprise for their mutual benefit for a long period of time. An international joint venture (IJV) is regarded as when one or more parents are headquartered outside the venture's country of operation, or if the JV has a significant level of operations in more than one nation (Geringer and Hebert, 1989). As a matter of fact, there are a number of typical JV approaches/structures, including (1) joint research and development; (2) joint distribution; (3) income access structure; (4) independent power project; and (5) management buy-out.

Some JV examples

Companies A and B enter into a JV for joint development hitherto carried out by A alone. Both A and B are able to gain development effort from the JV (Figure 8.7).

1   Shareholders' agreement;
2   Agreement for JV to purchase business from A;
3   Research and development agreements between (1) JV and A; and (2) JV and B.

Companies A and B enter into JV for exploiting in the territory of B a product range to which both participants contribute product and know how (Figure 8.8).

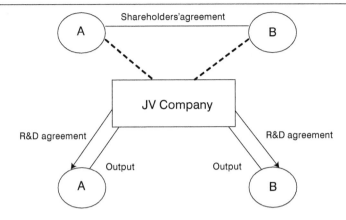

Figure 8.7 Joint research and development (adapted from Herzfeld and Wilson, 1996)

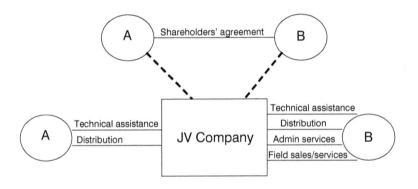

Figure 8.8 Joint distribution (adapted from Herzfeld and Wilson, 1996)

1   Shareholders' agreement;
2   Technical assistance and licence agreements between (1) A and JV; and
     (2) B and JV;
3   Distribution agreements between (1) A and JV and (2) B and JV;
4   Agreement for administrative services between B and JV;
5   Field sales and service agreement between B and JV.

Figure 8.9 shows the income access structure in which a Chinese parent
company and a French parent company form a UK JV to hold the business
currently carried on by the Chinese operating companies and the French
operating companies. UK JV holds all the voting shares in the Chinese
holding and the French holding, except that the Chinese parent and the
French parent retain one share each in their local companies. The income

access share entitles the holder to a dividend from the profits of the company in the holder's own tax jurisdiction, without routing it through the UK JV. The income share is affixed to the parent's holding in the UK JV so that the shares cannot be held by separate shareholders. This is achieved through a restriction on ownership in the constitutional documents of the Chinese holding and the French holding. Since it is the intention that the Chinese and French organizations should be run as one business, there may be provision for payments between the Chinese parent and the French parent to equalize their respective returns.

Figure 8.10 indicates the independent power project in which companies X, Y and Z (the sponsors) form a JV company to build, own and operate a power station in a host country and subscribe for equity share in the JV, and the JV enters into (1) a construction contract with the contractor to build the power station; (2) a power purchase agreement with a local utility under which the power purchaser agrees to purchase electricity at a price which enables the JV's debt to be serviced and produces an agreed return for the sponsors; (3) a fuel supply contract with a local fuel supplier; and (4) an operation and maintenance contract with an O&M contractor. The lenders agree to make loan financing available to the JV to carry out the project under a credit agreement. The JV assigns its entire interest in the power station. In addition, the government of the host country agrees to support the project under this agreement varying from an undertaking not to revoke a particular licence that the JV requires to a full guarantee of the power purchaser's obligations.

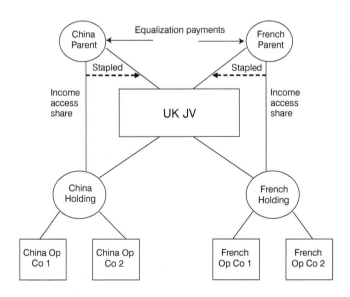

Figure 8.9 Income access structure (adapted from Herzfeld and Wilson, 1996)

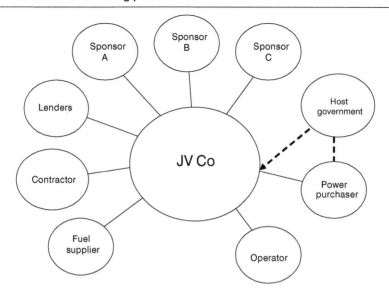

Figure 8.10 Independent power project (adapted from Herzfeld and Wilson, 1996)

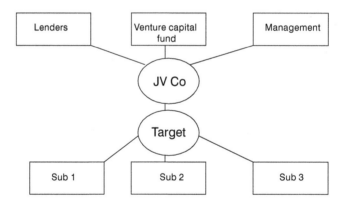

Figure 8.11 Management buy-out (adapted from Herzfeld and Wilson, 1996)

Figure 8.11 shows the management buy-out and it has the following features:

1   The senior management forms a special-purpose company to acquire the business of the target and subsidiaries;
2   The top management enters into long-term service contracts with the JV to ensure that their services continue to be available to the business;

3   The senior management and the venture capital fund enter into a subscription and shareholders' agreement under which each agrees to subscribe for an agreed number of shares in the JV;
4   The lenders consent to make loan financing available to the JV for the acquisition of the target under a credit agreement;
5   JV acquires the target under a share sale agreement from the vendor;
6   Each of the JV, target and its subsidiaries create security over their assets in favour of lenders.

JV process

Harrigan (2003) provided some guidelines for forming and managing a JV in which there are five important processes, including (1) finding suitable partners; (2) writing contracts; (3) deciding on ownership versus control; (4) using managers effectively; and (5) evaluating JV performance. Figure 8.12 shows the process of JV formation and management.

Hewitt (2001) considered some legal and transactional issues at the beginning of discussions regarding a proposed JV. The issues include (1) initial steps; (2) structure of JV; (3) local law issues; (4) contribution of

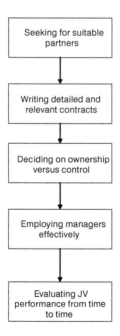

Figure 8.12 The process of JV formation and management

existing assets; (5) regulatory matters; (6) tax; (7) accounting; (8) employee issues; (9) intellectual property; and (10) property and related services.

Gibbons (1996) provided a checklist summarizing the basic steps in the deciding, preparing, negotiating and commencing phases of a Chinese JV project. The scopes of checklist are (1) eligibility; (2) preparation; and (3) start of negotiation to start of JV. In addition to these, there are some documents necessary for approval of a JV and some principal matters to be addressed in the feasibility report prior to approval of a JV.

Key elements for relational contracting success

To implement RC approach and process successfully and efficiently, there are two important tools: the relationship skilling workshop and group problem solving (Main Roads Project Delivery System, 2005).

Relationship skilling workshop (primarily for partnering and alliancing)

Purpose and objectives

Main Roads Project Delivery System (2005) stated that the purpose of conducting the relationship skilling workshop is to provide team members with effective team communication and interpersonal skills to help to maintain a continuous collaborative, 'best for project' approach to the project delivery. The objectives are to: (1) further develop an integrated team; (2) enhance interpersonal skills amongst team members; (3) develop motivators as specific to the project; and (4) develop techniques that can be used to resolve issues and problems that arise during the project.

Workshop content

1   Introduction and icebreakers. What is going well for all each individual both on the job and personally?
2   Moments of truth: the influence of positive and negative experiences, and levels of learning and service.
3   Exercise (levels of service).
    Each party notes down what they are currently doing that they think helps the other party and what they could do in the future.
    Each party notes what the other party is currently doing that is helpful to them and what they could do in the future.
    Comparison and discussion.
4   Motivation. What are the motivators?

5   Exercise. What each party is currently doing to motivate the team and what they could do in the future.
6   Conflict resolution. What influences behaviour and definitions of assertive, passive, aggressive and passive-aggressive behaviour.
7   Team dynamics. Theory and what level the team thinks it has reached.
8   Team exercise. Overview and recap what this process is to achieve.
9   Problem solving exercise in group synergy:
     Creative problem-solving techniques;
     Scenarios;
     Issues regarding the project;
     Decision making;
     Decision-making theory;
     Exercise in decision making;
     Ultimate team challenge.

Workshop participants

The relationship skilling workshop will typically involve the participants from the foundation workshop. This workshop, too, is usually 1.5 days with a dinner session.

Group problem solving

Main Roads Project Delivery System (2005) also viewed that group problem-solving processes form part of a strong client leadership a goes a long way towards gaining accepted project outcomes. It does this through enhanced integration of the supply chain in which there is increased participation by downstream suppliers in upstream processes, upstream suppliers in downstream processes, and end users in upstream processes. Correctly used, group problem-solving processes provide a client with better whole-life outcomes across the project life, from concept through to maintenance and use.

Main Roads Project Delivery System (2005) viewed that group problem-solving processes are structured, systematic and analytical; they allow a group of interested parties to optimize value in systems, processes, products and services. Value relates not only to price or value for the resources used but also to achieving what is of benefit or importance to the stakeholders in a particular circumstance. A major concept is that the collaborative output and thinking power of the group is greater than the sum of the outputs and thinking power of the individuals. Typical group problem-solving processes encompass participatory strategic planning, value management, value engineering, risk identification and management, partnering including RC and post-construction reviews.

Purpose and benefits

The group problem-solving processes aim to produce a range of alternative ideas which may be used by the client and other participants in the planning, design and construction process to help to take decisions about the project. The process may be used to arrive at a decision on a project or to develop alternatives for further analysis.

There are several benefits arising from group problem-solving processes, including (1) shared understanding among a wide range of stakeholders; (2) savings in life cycle costs; (3) a holistic solution to meet particular needs; (4) clarity, focus and improved communication; (5) savings in design and construction time; (6) reduced or well-managed risks; and (7) reflections and learning transferred to future projects.

Appropriate use

The use of group problem-solving processes varies with the project scale and complexity. Group problem solving can be used in different phases of a project and for different reasons:

1   In the early conceptual phase to identify the basic needs and the functional requirements to be incorporated into the project brief along with other physical characteristics;
2   At the preliminary design stage to ensure that the design options generated meet the functionality requirements, to eliminate factors that do not contribute to the functionality and to optimize the balance between function cost and worth;
3   In the detailed design phase to optimize the technical components of the project;
4   To identify and manage the risks associated with the project;
5   To identify the optimum delivery process and packaging for the project;
6   During any contract process to quickly identify and resolve issues that arise during the contract;
7   At project completion to identify specific learning arising from the project.

The most appropriate use of the various group solving processes is shown in Figure 8.13, which is also explained by the following seven main points:

1   Participatory strategic planning workshops may be used in the conceptual phase of a project;
2   Value management workshops may be used in the conceptual, planning and packaging decision phases;
3   Risk assessment and management workshops would be used from planning through to design;

4   Value engineering workshops would be used in the design and possibly the construction phase;

5   A post-construction review workshop would be held shortly after project completion;

6   Partnering workshops including team-building and team problem-solving techniques could be used at any stage of a project which involves a form of contractual relationship from the planning phase through to maintenance and operation;

7   In general, the participants at group problem-solving workshops will come from a more diverse background in the earlier phase of a project than the latter phases.

Group problem-solving process

Main Roads Project Delivery System (2005) stated that although the group problem-solving process differs slightly for the various techniques, underlying principles are similar. One of the major differences in the various workshops

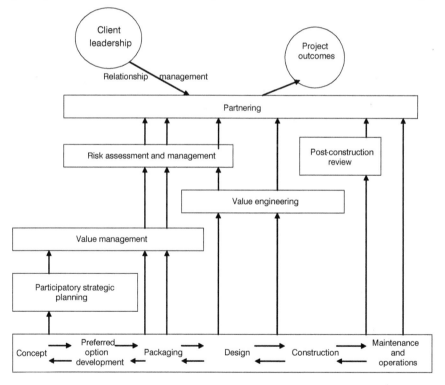

Figure 8.13 Appropriate use of the different group solving processes (reproduced from Main Roads Project Delivery System, 2005) (with permission for both print and online use from Department of Main Roads, Brisbane, Australia)

is the type of groups or individuals that are requested to participate in the workshops. The process starts by identifying the basic needs or project outcome required. The process is based on analysing the function performed by products or services. This involves identifying what they do and what they have to do at a high level to meet the project objectives. Once these high-level, project-functional objectives have been set, they are listed in priority order and ideas are generated as to how the functional objectives can be met.

Critical success factors

There are some factors that can assist in the success of the process (Main Roads Project Delivery System, 2005):

1   Dissemination of the information, implementation plans and final decisions to the participants of the workshop at the earliest opportunity;
2   Holding the workshop before strong opinions are generated regarding decisions;
3   If any non-negotiations have been decided, starting these at the beginning of the process;
4   Limit the issues to be addressed at a workshop to one key issue;
5   The process is best facilitated by trainers who are versed in group problem-solving techniques.

Chapter summary

This chapter begins by stating three major items to help make decisions for a suitable delivery strategy, including: (1) physical packaging of the project; (2) selection of an appropriate project delivery method; and (3) risk framework. Afterwards, it introduces the approaches and processes of different forms of RC. Finally, it emphasizes two important tools, the relationship skilling workshop and group problem solving, that can increase the efficiency of RC (mainly for partnering and alliancing) approaches and processes.

Chapter 9

Initial, interim and final workshops

Introduction

To procure a construction project through an RC approach successfully, it is important to carefully design the implementation process of the initial, interim and final RC workshops. A number of significant elements have been identified to make these workshops, mainly applicable to partnering and alliancing, a success.

The vital ingredients to make the initial RC workshop a success are:

1 Top management commitment from the principal stakeholder groups;
2 Implementation of smooth process of the initial RC workshop;
3 Joint workshop objectives;
4 Appropriate classification of workshop participants;
5 Selection of capable workshop facilitator(s);
6 Clear project outline;
7 Building up a charter through:
    a. RC objectives for core groups;
    b. framework for the charter;
    c. shared objectives;
    d. strengths of core groups; and
    e. potential hindrances and their recommendations;
8 Encouraging RC opportunities:
    a. at the creative phase;
    b. using brainstorming; and
    c. with cross-disciplinary grouping
9 Establishing communication and team-building skills;
10 Development of RC strategies;
11 Developing a suitable issue resolution ladder;
12 Developing a good performance monitoring mechanism for RC; and
13 Appropriate action plans.

On the other hand, the vital ingredients to make the interim RC workshop(s) a success are:

1   Implementation of smooth process of the interim workshop(s);
2   Joint workshop objectives;
3   Appropriate classification of workshop participants;
4   Training of capable in-house workshop facilitator(s);
5   Review of project status and performance;
6   Review of RC performance monitoring matrix;
7   Evaluation and identification of perceived benefits of RC;
8   Evaluation and identification of significant difficulties in implementing RC;
9   Evaluation of critical success factors for RC implementation;
10   Suggestion of remedial actions to overcome identified obstacles;
11   Review and update of the hierarchy framework for issue resolution; and
12   Re-nomination of RC champion.

Finally, the significant ingredients to make the final RC workshop success are:

1   Implementation of smooth process of the final workshop;
2   Joint workshop objectives;
3   Appropriate classification of workshop participants;
4   Overall review of project status and performance; and
5   Seeking opportunities for the future.

Significant elements to make the initial relational contracting workshop a success

Top management commitment from the principal stakeholder groups

Senior management executives from the principal stakeholder groups should deliver presentations at the beginning of the initial RC workshop. These presentations should underline the top management commitment of each contracting organization to make the RC project more successful through wholly embracing the philosophy and necessary actions implicit in RC.

Implementation of smooth process of the initial workshop

It is good for the facilitator to begin the initial RC workshop by stating an overview of the RC process and explaining the rationale behind it; it is suggested that the following items should be included during presentation:

1 Problems facing the construction industry;
2 Measures for improvement;
3 Define the RC approach: what is it and what is it not?
4 Global experiences of RC;
5 Local experiences of major local RC projects, useful KPIs and CSFs
6 How to implement RC, the process to be followed and the rundown
   of the initial RC workshop (Table 9.1).

Joint workshop objectives

It is suggested that the executives of the principal stakeholders should state the workshop objectives to develop understanding, agreement and a team commitment to a preliminary charter underpinned by clear, shared objectives with well-defined implementation strategies. The major objective of the initial RC workshop should align and unite all parties involved in the project with a shared goal of completing the work in a cost-effective and timely manner, which should be mutually satisfactory and beneficial. In fact, the initial RC workshop should aim to:

1 Build cooperation and mutual trust;
2 Align and unite all the parties involved;
3 Drive out waste and reduce cost;
4 Give greater programme certainty;
5 Improve relationships and reduce conflicts;
6 Improve quality and value in general;
7 Develop the framework of an RC charter;
8 Identify strategies to make RC work;
9 Propose a hierarchy of issue resolution mechanism;
10 Set up a regular RC performance monitoring system;
11 Sign the RC charter.

Appropriate classification of workshop participants

During the initial RC workshop implementation process, it is better to divide workshop participants into various groups by functional core group and cross-disciplinary group for discussion at various sessions. The functional core group should be based on the roles and responsibilities for each party, i.e. client, consultant, main contractor and subcontractor, while the cross-disciplinary group ought to be based on a mixture of representatives from different organizations involved in the project.

Table 9.1 Example of a rundown of a one-day initial RC workshop (adapted from Chan et al., 2006b)

Time	Activity	Action Parties	Duration
08:45 – 09:00	Registration	Host organization	15 min.
09:00 – 09:15	Welcome and Introduction	Host organization	15 min.
09:15 – 09:45	Overview of the RC process and rundown of the initial RC workshop	Facilitator	30 min.
09:45 – 09:55	Statements of commitment by the CEOs of the key organizations	CEOs of the key organizations	10 min.
09:55 – 10:05	Information phase: Brief description of the project	Project architect	10 min.
10:05 – 10:20	*Coffee break*	–	*15 min.*
10:20 – 10:40	Development of specific RC objectives of each organization	Break-out session (within core groups)	20 min.
10:40 – 11:10	Development of shared objectives	Whole team	30 min.
11:10 – 11:30	Development of the framework of a RC charter	Whole team	20 min.
11:30 – 12:15	Creative phase: Creating ideas to make RC work	Break-out session (cross-disciplinary teams)	45 min.
12:15 – 13:30	*Lunch*	–	*1 hr. 15 min.*
13:30 – 14:15	Creative phase: Creating ideas to make RC work	Break-out session (cross-disciplinary teams)	45 min.
14:15 – 14:45	Development of hierarchy framework for issue-resolution mechanism	Break-out session (cross-disciplinary teams)	30 min.
14:45 – 15:15	Monitoring RC performance	Facilitator	30 min.
15:15 – 15:30	Finalization of RC charter – preparation of action plan to follow key undecided issues during workshop	Whole team	15 min.
15:30 – 15:45	Collection of reactions of participants towards RC and the workshop	Whole team	15 min.
15:45 – 16:15	Final remarks	Senior management of the client	30 min.
16:15 – 16:45	*Coffee break*	–	*30 min.*

Selection of capable workshop facilitators

The selection of a competent, experienced facilitator would significantly affect the effectiveness and performance of RC. Therefore, the selection of the right facilitator is an essential ingredient to RC success. The stakeholders need to understand not only the potential benefits, but also training and support before starting the RC effort. It is vital that the facilitator remains involved throughout the whole life of the project. An independent qualified facilitator should be engaged for launching an initial RC workshop and assisting in fostering comfort and confidence with respect to the effective implementation of the RC process. A good facilitator should possess (American Arbitration Association, 1996):

1   A basic understanding of the construction industry;
2   Strong communication and listening skills;
3   Solid organizational and people skills;
4   Team-building skills;
5   Well-developed problem-solving and conflict management skills.

A good facilitator (American Arbitration Association, 1996):

1   Assists the group in focusing on common problems and goals;
2   Creates an environment of openness and trust;
3   Builds consensus and commitment on all topics discussed;
4   Establishes credibility and trust;
5   Matches the personality and style of the project stakeholders;
6   Maintains flexibility;
7   Controls the process;
8   Generates participation.

Clear project outline

In the information phase, the project architect should give a brief description of the development project in relation to the project status, design concept and project details. In addition, both the client's representative and the main contractor's representative should emphasize the client's expectations on the project, encompassing project completion with very high quality standards, on time, within budget, good teamwork, good communication, satisfaction by customers and end-users, and sound track-record of safety performance.

Building up a charter

There are some important steps to build up a charter, including (1) RC objectives for core groups; (2) framework for the charter; (3) shared objectives; (4) strengths of core groups; and (5) potential hindrances and their recommendations.

It is recommended that the workshop participants should first be categorized into functional core groups comprising client group, consultant group, main contractor group and subcontractor group. They should then be asked to list the individual specific objectives for each core group in this project using the 'goals and objectives' pro forma worksheet (see Appendix 1) with respect to the RC initiative (these should in no way conflict with the contractual commitments already in place). These in turn are presented to the whole team and it is apparent that there is a set of shared objectives across the groups.

The facilitator should write down keywords for each objective on a white board one by one and then they are regrouped to form a matrix summary. Based on a set of shared objectives identified by the functional core groups, the facilitator should summarize and consolidate them into various themes forming the basic framework for the RC charter.

The whole team should then examine the objectives of each core group and they should identify and agree with the consolidated themes. After identifying the shared objectives amongst the whole team, the facilitator should ask the participants to compile for their own core group the particular strengths that would be expected to affect the RC initiative using pro forma worksheet 2 (see Appendix 2). If the groups are able to identify these, then this can be found useful later in the development of the RC strategies.

After that, the facilitator should request the participants to list what each individual or organization believes will be the potential obstacles or hindrances which most likely impede the RC effectiveness using pro forma worksheet 3 (refer to Appendix 3). Furthermore, the participants should then be invited to suggest any recommendations to overcome those hindrances using pro forma sheet 4 (refer to Appendix 4). Each group will finally present their strengths, hindrances and recommendations to the whole team to gain consensus on their relevance.

Relational contracting opportunities

RC opportunities can be achieved through the creative phase, brainstorming techniques and cross-disciplinary grouping.

Creative phase

Creative phase allows the whole team to create new ideas that would help ensure a successful RC. The facilitator should first explain the concept of group brainstorming applied in this phase to all participants. In developing these ideas/opportunities, participants should be advised not to hold back on those ideas that might appear unrealistic as they could well prompt more acceptable ideas that are realistic. It is vital that judgement is suspended during this phase so that ideas are free flowing, creative and innovative.

Brainstorming technique

Brainstorming technique is then used and it is in fact a good technique to generate many ideas quickly within a group. That is where a group spontaneously tosses out ideas, regardless of how preliminary or far-fetched they may seem. Some principles of brainstorming are outlined below (Hale, 1996):

1 Suspend judgement. Criticism is ruled out, adverse judgement must be withheld until later, do not evaluate.
2 Be as wild as possible; the wilder the idea, the better. It is easier to tame down than to think up. Let your mind drift.
3 Generate as many ideas as possible. Pushing for quantity can assist in generating high-quality ideas.
4 Build and improve ideas. Suggest how ideas of others can be turned into better ideas and how two or more ideas can be joined into a better one.

Cross-disciplinary grouping

After brainstorming, the whole team should be further divided into cross-disciplinary teams, each of which should include at least one representative from each of the previous functional core groups. Each group will then be requested to try their best to generate as many ideas as possible for each RC objective (using pro forma sheet 2) (see Appendix 2). A short presentation by each of the groups should concisely summarize the key ideas or themes.

Establishment of communication and team-building skills

During the initial RC workshop, a key to RC success is to establish communication and team-building skills through team-building exercises. To do so, the following items should be noted, including: (1) getting to know each other by personality profiles; (2) learning empathy and listening skills; (3) setting forth individual authorities and responsibilities to avoid people passing the buck or ineffective communications; (4) delegating responsibilities as low as possible; and (5) identifying common goals.

Development into relational contracting strategies

Classification

Cross-disciplinary team grouping should be adopted to develop into RC strategies. The facilitator should explain the method for evaluating the brainstorming ideas developed previously. They should then be evaluated by the workshop team according to the classified possibilities as follows:

Realistic possibility: the idea has merit and identified as worthy of further evaluation during the workshop. The idea can be developed or combined with other ideas to develop an action, theme or strategy to benefit the RC objectives.

Remote possibility: the idea has some merit. However, it is remote and should be considered, if appropriate, outside the workshop.

Discarded: there is no benefit in further pursuing the idea and rejection of the idea is concurred by the group of the idea considered outside the scope of the workshop.

Development of strategies for implementation

After considering a range of ideas to assist in making the RC approach actually work, the participants should then concentrate on those ideas considered as being a realistic possibility.

The whole team should be involved in the explanations and discussions which support each idea. This provides a unified level of understanding and agreement of the merits of each idea which enables further development of strategies for implementation.

Cross-disciplinary groups are formed again with a mixture of company representation and hierarchies to ensure healthy interplays of views and knowledge of operations to discuss the classified ideas with respect to each theme determined. The participants should be invited to develop corresponding actions or implementation strategies based on those ideas identified as being a realistic possibility using pro forma sheet 5 (see Appendix 5). Further actions and strategies will be formulated by the team as a whole in accordance with the theme objectives determined by the framework for the RC charter (theme-objective-strategy format). The strategies from each group vary in presentation with some indicating a clear path of their development from the ideas through theme-objective-strategy while others provide their action recommendations directly from pro forma worksheet 2.

Development of a suitable issue resolution ladder

Enhanced communication networks and improved working relationships within the project team are crucial factors for successful RC. After the development of RC strategies, it is good for the facilitator to invite the whole team to put a proposal of a strategic model for issue resolution.

The facilitator should present a proposed framework for developing a model as the basis of resolving issues or problems, which is subsequently discussed by the team. It is expected that an issue resolution hierarchy will be established and agreed by the participants at the end of workshop. Then, the facilitator ought to ask them to assign suitable personnel of empowerment from each organization to make decisions at each operational level ranging from frontline site staff to senior management. Each organization needs to fill in a specific form using pro forma worksheet 6 (see Appendix 6) for issue elevation, and the names of the assigned persons are finally concurred by the team as a whole.

The following essential principles for issue resolution (Skues, 1996) are recommended to be agreed and adopted for the project:

1   Issues are to be resolved at the lowest level;
2   There should be a simple hierarchy of communication levels;
3   Not all RC members have to be involved at each level or in every issue;
4   Any issues should be given a maximum of two attempts to be resolved at a particular level;
5   Any unresolved issues must be elevated to the next level for resolution;
6   Resolve within an agreed timescale;
7   Record outcome.

Development of a good performance-monitoring mechanism for relational contracting

After developing the issue elevation hierarchy, it is crucial to regularly monitor the RC performance of the project against the objectives of the RC charter. An opportunity to review and comment regularly is better to be provided on each party's perceptions of the success or otherwise of the RC initiatives. It is of utmost importance that this review does not go astray within the agenda of other necessary meetings and is justified as a separate and regular review meeting.

It is good for the facilitator to propose a mechanism matrix for monitoring the performance of RC and necessary modifications to gain consensus of the team as a useful evaluative tool to record feedback on the RC initiatives. Each of the shared (common) objectives will be listed with theme headings indicating relative assessment of performance from 1 (very unsatisfactory) up to 5 (very satisfactory) tracking over a number of time periods, e.g. on a monthly basis. A copy of the RC performance monitoring matrix is attached in Appendix 7.

The matrix can be used to score the effectiveness of the team as a whole, one's own organization or another partner's performance. It is essential to monitor the trends of RC performance on a monthly or quarterly basis. This will generate useful pointers into the perceptions of RC success and will assist in identifying further efforts or initiatives.

Individual sheets should be delivered to each organization or subcontracting party well before the monthly RC meetings such that organizations can score themselves first. These individual scores are then compared to the average score of the whole project team at the meetings. Apparent discrepancies or low scores can further be discussed and analysed for remedial actions.

Continuous monitoring programme

A steering group in the form of RC champions is suggested to set up and implement the appropriate action plans as agreed in the workshop and to evaluate RC performance of the project. The role of an RC champion is to:

1   Circulate individual evaluation forms to the project team monthly for completion;
2   Collect completed evaluation forms for compilation;
3   Review the results and take the necessary corrective measures that are appropriate.

The RC champions for each of the functional core groups involved in this project should be nominated by the whole team. It is stressed by the facilitator that RC itself would not generate a successful outcome on individual basis. The objective of RC is to ensure that all signatories to the RC charter would strictly adhere to the pledge, and if not, to find out reasons behind and to take corrective actions later.

Appropriate action plans

After establishing a mechanism for monitoring the RC performance of the project, the facilitator should once again present the goals, shared objectives and draft RC charter to the participants to enable a finalization of the charter. The facilitator should also review the objectives of the workshop and demonstrate the achievement through the structured programme that has been followed in the workshop.

Finally, the facilitator should initiate appropriate action plans to the participants to address the key issues that may not be concluded during the workshop but can be resolved by members of the project team subsequently. It includes setting a date to sign the RC charter. For each action in the plan, a specific person should be assigned responsibility to ensure completion of the task within a specified date. In fact, a coordinator or RC monitoring group should be assigned to follow up the implementation of action plan.

Significant factors to make the interim relational contracting workshop a success

Smoothing the process of the interim relational contracting workshop(s)

It is good for senior management executives from both the client and the main contractor to present briefly at the beginning of the interim RC workshop(s). These presentations should re-emphasize senior management's commitment to the RC project through wholly embracing the philosophy and essential actions implicit in RC.

Then, the facilitator should start the workshop by providing an overview of the RC process and the programme rundown. The key elements of the RC charter should be further restated to project participants. It is good to follow that with a brief review of the project status from various contracting parties so that the newcomers will also know the foundation of RC principles. After that, the RC monitoring matrix should be reviewed to evaluate the RC performance of the project. The workshop participants are suggested to be divided into cross-disciplinary teams and they are invited to fill in some tailor-made pro forma worksheets before having their discussions.

The following RC dimensions are suggested to be discussed:

1 Evaluation and identification of perceived benefits of RC;
2 Evaluation and identification of significant difficulties in implementing RC;
3 Evaluation of critical success factors for RC implementation;
4 Suggestion of remedial actions to overcome identified obstacles.

After peer discussion, it is good for each group to nominate a representative to report their findings. Then, the hierarchy framework for issue resolution mechanism ought to be reviewed and updated, and the RC champion should be re-nominated. Table 9.2 shows an example of a rundown of an interim RC workshop.

Joint workshop objectives

The primary objective of the interim RC workshop is to strengthen the cooperative nature of the project based on commitment, equality and trust amongst all the project team members. In essence, it is essential to structure the interim RC workshop to achieve the following objectives:

Table 9.2 Example of a rundown of an interim RC workshop (adapted from Chan et al., 2004f)

Time	Activity	Action Parties	Duration
09:00 – 09:30	Registration/Greeting and refreshments	Host organization	30 min.
09:30 – 09:45	Welcome and introduction	Host organization	15 min.
09:45 – 10:15	Overview of the RC process and rundown of the interim RC workshop	Facilitator	30 min.
10:15 – 10:30	Restate the key elements of RC charter towards project participants	Facilitator	15 min.
10:30 – 10:45	*Coffee break*	–	*15 min*
10:45 – 11:30	Brief review of project status and performance	Various parties	45 min.
11:30 – 12:15	Review the RC monitoring matrix to evaluate the RC performance of the project (e.g. explore the reasons for low performance scores)	Whole team	45 min.
12:15 – 13:30	*Lunch*	–	*1 hr. 15 min.*
13:30 – 14:15	Evaluate and identify perceived benefits of RC	Break-out session (cross-disciplinary teams)	45 min.
14:15 – 15:00	Evaluate and identify significant difficulties in implementing RC	Break-out session (cross-disciplinary teams)	45 min.
15:00 – 15:45	Evaluate critical success factors for RC implementation and suggest remedial actions to overcome identified obstacles	Break-out session (cross-disciplinary teams)	45 min.
15:45 – 16:00	*Coffee break*	–	*15 min.*
16:00 – 16:30	Presentation of cross-disciplinary teams	Break-out session (cross-disciplinary teams)	30 min.
16:30 – 17:00	Review and update the hierarchy framework for issue escalation mechanism	Whole team	30 min.
17:00 – 17:15	Nominate a new group of RC champions from project participants	Whole team	15 min.
17:15 – 17:30	Seek responses of participants to RC and the interim workshop	Whole team	15 min.
17:30 – 17:45	Final remarks	Host organization	15 min.

1 Introduce the concept of RC to new project stakeholders;
2 Review the RC monitoring matrix to evaluate the RC performance for the project;
3 Identify reasons for low performance and recommend remedial action;
4 Identify the perceived benefits and critical success factors for RC implementation;
5 Identify the major obstacles to RC success and their corresponding corrective actions;
6 Reiterate the key elements of the RC charter;
7 Review and update the hierarchy framework for issue escalation mechanism;
8 Nominate a new group of RC champions from the project participants.

Appropriate classification of workshop participants

During the interim RC workshop implementation process, it is better to divide workshop participants into various groups by cross-disciplinary group for further discussion at various sessions. As mentioned previously, the cross-disciplinary group ought to be based on a mixture of representatives from different organizations involved in the project.

Training of capable in-house facilitators

In addition to the external facilitators, in-house facilitators should be trained to run the interim RC workshop. In fact, to do so it may help to promote the RC concept at the site level and to junior staff so that it can help to reap the fullest benefits of RC. It should be noted that like external facilitators, the in-house facilitators should possess:

1 A sound understanding of the construction industry;
2 Strong communication and listening skills;
3 Solid organizational and interpersonal skills;
4 Demonstrable team-building skills;
5 Well-developed problem-solving and conflict-management skills.

And he/she should:

1 Assist the group to focus on common problems and goals;
2 Create an environment of openness and trust;
3 Build consensus and commitment on all topics discussed;
4 Develop credibility and trust;
5 Match the personality and style of the project stakeholders;
6 Maintain flexibility;
7 Control the process;
8 Generate participation.

Review of project status and performance

After restating the key elements of RC to project participants, the facilitator should review project status and performance. Typical items to be reviewed include: (1) progress updates; (2) cost performance; (3) quality issues; and (4) safety performance.

Review of the relational contracting performance monitoring matrix

Then, the RC performance monitoring matrix should be presented and reviewed to evaluate the RC performance of the project. It should be noted that the whole team should explore the reasons for KPIs with low performance scores. For example, investigation should be conducted if there is a sudden drop in the safety culture for a short period of time.

Evaluation and identification of perceived benefits of relational contracting

Each of the cross-disciplinary groups should write down the perceived benefits of RC in the project using pro forma worksheet 7 (see Appendix 8) via brainstorming discussions. A number of benefits of RC identified from academic literature can then be used as a basis of pro forma sheet 8 (Appendix 9). Participants will then be requested to rate these benefits according to their experience on a five-point Likert scale (from Not Significant to Extremely Significant). The empirical findings can then be compared with the perceived benefits of RC identified during the brainstorming discussions to see if they are consistent.

Evaluation and identification of significant difficulties in relational contracting

Each of the cross-disciplinary groups can be further requested to brainstorm the major difficulties encountered in adopting RC approach in the project with the aid of pro forma sheet 9 (see Appendix 10). Then, a number of major difficulties in implementing RC identified from academic literature can be used as a basis of pro forma 10 (Appendix 11). Similarly, participants will then be requested to rate these major difficulties according to their experience on a five-point Likert scale (from Not Significant to Extremely Significant). The empirical findings can then be compared with the perceived major difficulties in implementing RC identified during the brainstorming discussions to see if they are consistent.

Evaluation of critical success factors for implementing relational contracting

Each of the cross-disciplinary groups can be further requested to suggest and write down the CSFs for adopting RC in the project using pro forma sheet 11 (see Appendix 12) via brainstorming discussions. A number of CSFs for adopting RC identified from academic literature can then be used as a basis of pro forma worksheet 12 (Appendix 13). Participants will then be requested to rate these CSFs according to their experience on a five-point Likert scale (from Not Significant to Extremely Significant). Similarly, the empirical findings can then be compared with the perceived CSFs of RC identified during the brainstorming discussions to see if they are consistent.

Suggestion of remedial actions to overcome identified obstacles

Each of the cross-disciplinary groups can be further requested to formulate strategies to overcome the major difficulties in implementing RC with the aid of pro forma worksheet 13 (see Appendix 14).

Review and update of the hierarchy framework for issue resolution

The hierarchy framework for the issue escalation mechanism should be reviewed. The updated issue escalation ladder should continue to set out the levels ranging from front-line site staff to senior management. Names of the corresponding personnel from each organization should be recorded in the escalation ladder.

Re-nomination of relational contracting champions

It should be agreed unanimously to re-nominate the RC champions to keep the RC spirit alive.

Appropriate action plans

Finally, the facilitator should initiate appropriate action plans to the participants to address the key issues that may not be concluded during the workshop but can be resolved by members of the project team subsequently. For each action in the action plan, a named person should be assigned responsibility to ensure completion of the task within a specified date. In fact, a coordinator or RC monitoring group should be assigned to follow up the implementation of action plans.

Significant factors to make the final relational contracting workshop a success

Smoothing the process of the final relational contracting workshop

It is good for the facilitator to start the final workshop by providing an overview of the RC process and the programme rundown. A typical final RC workshop should include: (1) introduction (workshop objectives and review charter); (2) history of the project; (3) sharing of partnering champions; (4) learning from champions' report; (5) quick review of partnering performance; and (6) seeking opportunities for the future.

Joint workshop objectives

The major objectives of the final RC workshop should: (1) review the overall partnering performance of a project; (2) summarize lessons learned from the whole partnering journey; and (3) seek opportunities for the future.

Appropriate classification of workshop participants

During the final RC workshop implementation process, it is better to divide workshop participants into different groups by cross-disciplinary group for further discussion at various sessions. As mentioned previously, the cross-disciplinary group ought to be based on a mixture of representatives from different organizations involved in the project.

Overall review of project status and performance

After reviewing the history of the project and sharing of partnering champions, the overall project status and performance should then be reviewed. Typical items to be reviewed include: (1) progress updates; (2) cost performance; (3) quality issues; and (4) safety performance.

Seeking opportunities for the future

Having reviewed the overall project status and performance, the facilitator should invite participants to seek opportunities for the future, such as brainstorming the short- and long-term partnering strategies, and brainstorming to apply the partnering approach to future projects.

Chapter summary

This chapter proposes some guidelines to implement initial, interim and final RC (primarily applicable to partnering and alliancing) workshops successfully. Significant elements are identified as crucial to make the initial, interim and final RC workshops successful.

The significant factors for making the initial RC workshop successful are: (1) top management commitment from the principal stakeholder groups; (2) implementation of smooth process (rundown) of the initial RC workshop; (3) joint workshop objectives; (4) appropriate classification of workshop participants; (5) selection of capable workshop facilitator(s); (6) clear project outline; (7) building up a charter through (a) RC objectives for core groups; (b) framework for the charter; (c) shared objectives; (d) strengths of core groups; and (e) potential hindrances and their recommendations; (8) RC opportunities by (a) creative phase; (b) brainstorming technique; and (c) cross-disciplinary grouping; (9) establishing communication and team building skills; (10) development into RC strategies; (11) developing a suitable issue resolution ladder; (12) developing a good performance monitoring mechanism for RC; and (13) appropriate action plans.

The vital elements for making the interim RC workshop(s) successful are: (1) implementation of smooth process (rundown) of the interim workshop(s); (2) joint workshop objectives; (3) appropriate classification of workshop participants; (4) training of capable in-house workshop facilitator(s); (5) review of project status and performance; (6) review of RC performance monitoring matrix; (7) evaluation and identification of perceived benefits of RC; (8) evaluation and identification of significant difficulties in implementing RC; (9) evaluation of critical success factors for RC implementation; (10) suggestion of remedial actions to overcome identified obstacles; (11) review and update of the hierarchy framework for issue resolution; and (12) re-nomination of RC champion.

Finally, the significant factors to make the final RC workshop successful are: (1) implementation of smooth process (rundown) of the final workshop; (2) joint workshop objectives; (3) appropriate classification of workshop participants; (4) overall review of project status and performance; and (5) seeking opportunities for the future. In Chapter Ten, recommendations for developing a best practice framework for different forms of RC will be explored.

Chapter 10

Recommendations for developing a best practice framework for different forms of relational contracting projects

Introduction

After identifying a number of CSFs for different forms of RC projects and depicting various forms of RC approach and process, some recommendations are made in this chapter to develop a best practice framework for various forms of RC projects. This will form the basic guidelines for implementing successful RC with different forms for future construction projects.

For the partnering approach and process, the following recommendations and CSFs are made:

- Structured partnering process and careful design of activity contents;
- Early implementation of partnering process;
- Regular monitoring of partnering process;
- Selection of qualified facilitators for partnering workshops;
- Training of in-house facilitators;
- Appointment and true empowerment of the partnering champions.

The following six CSFs should be noted before implementing the partnering approach:

- Mutual trust amongst project participants;
- Partnering spirit of win-win attitude;
- Long-term commitment;
- Top management support;
- Effective communication;
- Establishment of conflict resolution strategy.

For alliancing approach and process, the following recommendations and CSFs are made:

- Proper set up of alliance framework;
- Selection of right participants to join the alliance.

The following four CSFs should be noted before implementing the alliancing approach:

- Equitable sharing of risk amongst different parties;
- Mutual trust amongst project participants;
- Long-term commitment;
- Efficient cooperation and communication.

For the PPP approach and process, the following recommendations and CSFs are made:

- Conducting a feasibility study;
- Community endorsement of a PPP project proposal.

The following seven CSFs should be noted before implementing the PPP approach:

- Appropriate allocation of risk;
- A strong private consortium;
- Available financial market;
- Adequate legal framework;
- Local government support;
- Stable political and social environment;
- Transparent and efficient procurement.

For the JV approach and process, the following recommendations and CSFs are made:

- Finding suitable partners;
- Deciding on ownership structure of JV.

The following four CSFs should be noted before implementing the JV approach:

- Mutual trust amongst project participants;
- Long-term commitment;
- Top management support;
- Efficient coordination and cooperation.

Recommendations to develop a best practice framework for different forms of relational contracting projects

Partnering approach and process

Structured partnering process and careful design of activity contents

A structured partnering process helps to facilitate the partnering spirit more efficiently. It can include one preliminary meeting, one initial partnering workshop, regular champions meetings, a number of interim partnering workshops and one final partnering workshop.

The activity contents for initial partnering workshop can include: (1) common goals and objectives; (2) values of partnering; (3) critical success factors; (4) partnering process to subcontractors; (5) partners' strengths; (6) obstacles to success; (7) ideas on how to overcome obstacles; (8) ideas to make partnering work; (9) partnering charter; (10) action plan; (11) game; (12) problems resolution process; and (13) nomination of partnering champions.

The activity contents for interim partnering workshops can encompass: (1) review performance measures; (2) revision of issue escalation ladder; (3) revision of partnering strategies; (4) discussion of the identified waste and improvement areas; (5) subgroup summaries of major obstacles; (6) corrective action/resolution/opportunity; (7) game; and (8) re-nomination of partnering champions.

The activity contents for final partnering workshop can comprise: (1) overall review; (2) sharing of experience; and (3) opportunity for the future.

Early implementation of partnering process

It is commonly accepted that the earlier the implementation of partnering process, the more successful the project will be. It takes time for project team members to tune into the partnering spirit. An early implementation will enhance the likelihood of maximizing the benefits of partnering, both tangibly and intangibly.

Regular monitoring of partnering process

Regular interim workshops are effective to strengthen the partnering spirit of all parties over the life of project. The joint evaluation of project progress is one of the key outcomes of interim partnering workshops. The aims of interim workshops are to review performance measures, revise the issue escalation ladder, revise partnering strategies, discuss the identified waste and improvement areas, and take corrective action, resolution and opportunity.

Measurable goals can be used to determine and evaluate individual progress performance.

It is advisable to continue with an independent facilitator or experienced champion from one of the contracting parties to chair the first few review workshops. Once the process is established and the team champions gain adequate experience and confidence, it can then be handed over to them for their continued performance monitoring (APM-HK, 2003).

Selection of qualified facilitators for partnering workshops

The selection of a competent, experienced facilitator would significantly affect the effectiveness and performance of partnering (Chow, 2002). Therefore, the selection of the right facilitator is an essential ingredient to partnering success. The stakeholders need to understand not only the potential benefits, but also training and support prior to starting the partnering effort. It is vital that the facilitator remains involved throughout the whole life of the project. An independent qualified facilitator can be engaged for launching an initial partnering workshop and assisting in fostering comfort and confidence with respect to the effective implementation of the partnering process (American Arbitration Association, 1996).

Training of in-house facilitators

In addition to the external facilitators, in-house facilitators can be trained to promote the partnering concept to the site level and junior staff. This can ensure that participants can understand the partnering concept fully prior to the actual implementation. Adequately trained in-house facilitators may be considered for the subsequent interim/review partnering workshops.

Appointment and empowerment of the partnering champions

The primary roles of the partnering champions are to foster the partnering concept and assist those who might slip back to their traditional ways to maintain the partnering spirit. The appointment of partnering champions is to ensure that partnering principles do not slip out of focus and partnering process can be monitored regularly. In fact, the partnering champions act as a steering group to develop and implement the action plans following the partnering workshops. Individual evaluation forms can be circulated to the project team regularly for completion and returned to the nominated champions for review. The team then reviews the results and takes the necessary corrective measures. For these reasons, partnering champions should be truly empowered by their senior management to implement partnering efficiently and effectively.

Awareness of critical success factors for partnering

To ensure successful partnering implementation, the following CSFs are suggested.

Mutual trust amongst the project participants

For partnering to work, it is necessary for parties to have mutual trust towards other partners. They should have the belief that others are reliable in an exchange relationship. It is essential to 'open' the boundaries of the relationship as it can relieve stress and enhance adaptability, and encourage information exchange and joint problem solving and promise better outcomes (Mohr and Spekman, 1994; Cheng et al., 2000, Bayramoglu, 2001).

Partnering spirit of win-win attitude

As partners work together towards a common goal, each party agrees to examine each situation and strive to attain a win-win solution (Slater, 1998; Lazar, 2000).

Long-term commitment

Long-term commitment can be regarded as the willingness of the involved parties to integrate continuously (Bresnen and Marshall, 2000a; Cheng et al., 2000). More committed parties are expected to balance the attainment of short-term objectives with long-term goals, and achieve both individual and joint missions without raising the fear of opportunistic behaviour (Mohr and Spekman, 1994; Romancik, 1995).

Top management support

Commitment and support from top management is always a key prerequisite for a successful partnering venture. As senior management formulates the strategy and direction of business activities, their full support and commitment are critical in initiating and leading the partnering spirit (Cheng et al., 2000).

Effective communication

Partnering requires timely communication of information and the maintenance of open, direct lines of communication amongst all project team members. For construction projects, problems need to be solved at the lowest possible level (on site) immediately. If it is only used for routine matters while important issues are conveyed from each site office back to the respective head offices and then back to the site office before any interactions, partnering will fail (Moore et al., 1992). It is clear that effective communication skills can help to facilitate exchange of ideas, visions and overcome difficulties (Cheng et al., 2000).

Establishment of conflict resolution strategy

Because of the discrepancy in goals and expectations, conflicting issues are commonly observed to exist among parties. Conflict resolution techniques like coercion and confrontation are counter-productive and fail to reach a win-win situation (Lazar, 2000). In fact, the conflicting parties look for a mutually satisfactory solution and this can be achieved by joint problem solving to seek alternatives for the problematic issues. Such a high level of participation among parties may help them to secure a commitment to the mutually agreed solution (Cheng et al., 2000).

Alliancing approach and process

Proper set up of alliance framework

To implement alliancing successfully, it is essential to set up an alliance framework properly. As mentioned in Chapter Eight, project alliancing is especially suitable for projects under the following situations (Department of Treasury and Finance, 2006): (1) a number of complex and unpredictable risks with complicated interfaces; (2) complex stakeholder issues; (3) complicated external opportunities and threats that can only be managed successfully with collective efforts; (4) very tight schedule; and (5) output specifications are unclear. In fact, it is helpful to confirm the appropriateness of the alliance through three steps: (1) primary tests; (2) analysis of procurement methods against project objectives; and (3) supplementary comparative assessments. Having decided to use an alliance approach to procure a construction project, it is important to develop the alliance framework with the following principles (Main Roads Project Delivery System, 2005): (1) collective responsibility for performance with an equitable risk and reward sharing; (2) a primary emphasis on real gain-pain sharing business outcomes; (3) a peer relationship with equal say; (4) all decisions are best for the project; (4) clear responsibilities within a no-blame working culture; (5) fully open book transactions; and (6) open and honest communication.

Selection of right participants to join the alliance

The selection of the right participants to join the alliance is crucial to a successful alliance (Department of Treasury and Finance, 2006). To ensure that the most suitable alliance team is selected, a well-experienced selection panel together with a rigorous selection process should be established. In principle, a rigorous selection process should at least encompass: (1) initial short-listing; (2) interviews/final short-listing; and (3) workshops to select the most preferred.

Awareness of critical success factors for alliancing

To ensure a successful alliance, the following CSFs are suggested.

Equitable sharing of risk amongst different parties

Risks that are transferred to contractors under a traditional procurement method should be shared fairly amongst different parties under an alliance contract (Manley and Hampson, 2000). In particular, it is vital to share financial risks equitably between the alliance participants so that they can either all gain or all lose together financially depending on team performance.

Mutual trust amongst project participants

Like partnering, it is essential for parties to have mutual trust with other partners so as to achieve alliance success. If trust is developed between the collaborating parties, the alliance is very likely to be successful. In fact, trust normally comes from the reputation of the trustee organizations and their capacity to adhere to the commitment (Xu et al., 2005).

Long-term commitment

A successful alliance requires commitment by different parties to achieve common goals (Thorpe and Dugdale, 2004). Rowlinson and Cheung (2004c) observed that commitment has a strong impact on the team and alliance culture, indicating that alliance has a high chance of success when there is a sufficient long-term commitment from top management.

Efficient cooperation and communication

Efficient cooperation and communication amongst project team members is essential for successful relationship building under an alliance contract (Walker et al., 2000b). The intense integration of alliance partners through the whole collaborative process requires excellence in communication at a personal level, at a business level and at an operational level (Hauck et al., 2004; Walker et al., 2002).

Public-private partnership approach and process

Launch of a feasibility study

To implement a PPP successfully, it is vital to conduct a feasibility study before deciding to adopt a PPP approach to procure a construction project. Hong Kong Efficiency Unit (2003) stated that a decision to proceed with the PPP approach following from the feasibility study recommendations should address three items: (1) provision of service specifications, which can be agreed upon by stakeholders and tested in the market; (2) assessment of the financial viability of the project; and (3) investigation of the possibility for a client to have essential legal powers to let successfully a contract for the provision of the service.

Community endorsement of a public-private partnership project proposal

Having conducted the feasibility study, the client can conduct technical assessments before inviting proposals. Like any project, there is a need to ensure that those individuals or bodies who are directly interested in the PPP project are kept fully informed and properly consulted at all stages of the project. Therefore, the community endorsement of a PPP project proposal can ensure the political and social stability to implement the PPP project (Hong Kong Efficiency Unit, 2003).

Awareness of critical success factors for public-private partnership

To ensure successful PPP implementation, the following CSFs are suggested.

Appropriate allocation of risk

Appropriate risk allocation and risk sharing is crucial for PPP success (Qiao et al., 2001; Li et al., 2005). In general, risk should be allocated to the party best able to manage it, and this helps to reduce individual risk premiums and the overall cost of the project because the party can manage a particular risk under the best situation.

A strong private consortium

Many researchers have identified a strong private consortium as a CSF for PPP projects (Jefferies et al., 2002; Tiong, 1996; Birnie, 1999; Li et al., 2005). Before contracting out the PPP projects, the government needs to ensure that the parties in the private sector consortium are sufficiently competent and financially capable of taking up the projects (Asian Development Bank,

2007). It is recommended that it is better for private companies to explore other participants' strengths and weaknesses and, when deemed appropriate, join together to form consortia capable of synergizing and exploiting their individual strengths.

Available financial market

For a PPP project to be successful, the private contractor/concessionaire can easily access a financial market with the associated benefits of lower financial costs (Qiao et al., 2001; Jefferies et al., 2002; Akintoye et al., 2001; Li et al., 2005). In fact, an accessible financial market is an incentive for private sector to take part in PPP projects. Asian Development Bank (2007) stated that project financing is a critical factor for private sector investment in public infrastructure projects. It would be an incentive for private sectors to take up PPP projects when there is an efficient and mature financial market to benefit low financing costs and diversified range of financial products.

Adequate legal framework

It is believed that an independent, equitable, efficient and favourable legal framework is an important factor for successful PPP project implementation (Qiao et al., 2001; Jefferies et al., 2002; McCarthy and Tiong, 1991; Akintoye et al., 2001; Asian Development Bank, 2007). Adequate legal resources at reasonable costs should be available to deal with the amount of legal structuring and documentation required. A transparent and stable legal framework can help to make the contracts and agreements bankable. An adequate dispute resolution system should be able to help the stability of the PPP arrangements.

Local government support

Under PPP contracts, the government should endeavour to procure the assets and to deliver the services on schedule with high quality and meet the agreed service benchmarks throughout the life of the contract (Asian Development Bank, 2007). The government should be given the primary role of industry and service regulation. It should also be flexible in adopting innovations and new technology, and provide strong support and make incentive payments to the private sector where appropriate.

Stable political and social environment

A stable political and social environment is a prerequisite for a successful PPP (Qiao et al., 2001; Zhang et al., 1998; Asian Development Bank, 2007). This in turn relies on the stability and capability of the host government. In fact, political and social issues go beyond private sector domain and they should be handled by the government.

Transparent and efficient procurement

For the sake of implementing PPP effectively, it is necessary to have a transparent and efficient procurement process (Jefferies et al., 1992; Kopp, 1997; Gentry and Fernandez, 1997; Li et al., 2005). Asian Development Bank (2007) viewed that clear project brief and client requirement can assist in achieving these in a bidding process.

Joint venture approach and process

Search for suitable partners

There is an increasing trend for firms to seek partnerships to stay ahead of the competition in the modern global economy (Yan and Luo, 2001). Multiple forces and incentives exist for companies to create international joint ventures and alliances. In fact, partner selection is crucial for JV success. Appropriate selection of a local partner can improve a multinational firm's disadvantage of 'foreignness'. For instance, a preferred local partner can make it possible to invest in industries that are subject to local government restriction against direct foreign investment, and can assist a multinational firm to gain access to marketing and distribution channels that are available only to local business. On the other hand, to the local firm, selection of a suitable foreign partner is also crucial because a desirable multinational partner can bring to the JV advanced technologies, know-how, management expertise and international channels to the marketplace (Yan and Luo, 2001).

Decision on ownership structure

Structure of JV ownership has a strong effect on risk sharing and resource commitment of the partners and the vulnerability and strategic flexibility of the JV (Yan and Luo, 2001). In fact, ownership structure of a JV has critical and direct implications for the JV's operation (Yan and Luo, 2001). First, it is a common practice that a JV's ownership structure is associated with the JV's profit remittance scheme. Second, ownership structure reflects a partner firm's investment strategy. This is because the majority or minority holdings in a JV are closely linked to the firm's capability to contribute strategic resources to the partnership and to the firm's established business practices. Third, ownership structure is expected to have a significant effect on the structure of control over the venture's operations. Fourth, a JV's relative strength within an interdependent, multinational network can reduce its vulnerability to host government.

Awareness of critical success factors for a joint venture

To ensure successful JV implementation, the following CSFs are suggested.

Mutual trust amongst project participants

The need of trust between partners in a JV project has been identified as an important element of a long-term joint venture relationship (Madhok, 1995). Adnan and Morledge (2003) agreed that trust is a significant ingredient of managing relationships, and within organizations, trust leads to more effective implementation of strategy, greater managerial coordination and more effective team works.

Long-term commitment

Adnan and Morledge (2003) opined that commitment reflects the actions of some key decision makers regarding continuation of the relationship, acceptance of the common goals, the values of the partnership, and the willingness to invest resources in the relationship. Srindharan (1995) agreed that commitment is vital because it provides a long-term basis, resources and capabilities to the specific needs of the JV for its success.

Top management support

Like other types of RC projects, support from senior management is crucial for JV success.

Efficient coordination and cooperation

It is thought that cooperative behaviour between firms can assist in reducing potentially burdensome monitoring and safeguards costs within a JV project (Adnan and Morledge, 2003).

Chapter summary

This chapter has made some recommendations for developing a best practice framework for different forms of RC implementation. This will form the basic guidelines and skeleton for implementing successful RC with various forms in future construction projects. In general, the recommendations are divided into two main categories. The first category is related to different forms of RC approach and process and the second category is the awareness of CSFs for various forms of RC projects. In Chapter Eleven, case studies in partnering will be presented and analysed.

Part III
Case studies in relational contracting

Chapter 11

Case studies in partnering

Introduction

This chapter shows how partnering is implemented in four different nations. A total of 16 case studies in partnering adopted in the USA, the UK, Australia and Hong Kong are selected for demonstration purposes.

Case study 1

Interchange improvement – A case study of a construction project in the USA (extracted from the website of the American Association of State Highway and Transportation Officials (AASHTO): http://cms.transportation. org/?siteid=38&pageid=377) (permission has been obtained for both print and online use from the AASHTO).

Background

This project was to reconstruct an existing diamond interchange to alter it to a single-point urban interchange. The client was Maryland State Highway Administration, the designer was URS Corporation, and the main contractor was Corman Construction Inc. During the reconstruction period, all parties used partnering techniques to identify and resolve issues when they arose. The reduction in the number of construction phases was the most crucial target for them to achieve. Originally, three construction phases were planned, but due to the effective partnering approach, the second and third phases were combined. By doing so, it increased construction efficiency and still maintained all vehicular and pedestrian movement. In addition, by combining these two phases, a spilt traffic phase was reduced by constructing a temporary cantilever sidewalk and installing temporary support beams and temporary bridge piers under the construction joint to permit more of the phase one deck to be used by the traveling public.

Major challenges

A number of major challenges were encountered in this project. First, it was constructed in a highly congested area and needed to be completed quickly. It had to maintain all pedestrian and vehicular traffic movements across the bridge during the whole life of the project. From the contractor's point of view, the contract duration needed to be decreased to take advantage of the contract's early completion incentive clause. From the designer's and owner's perspectives, it was vital to complete the project as quickly as possible without sacrificing safety or design criteria whilst maintaining a high quality of work. Another challenge for all team members was satisfying the design criteria and requiring all traffic movements to be maintained.

Partnering principles

Prior to the start of the project, a partnering workshop was held and a partnering charter was created. Monthly partnering meetings were then held to identify issues and track the methods that could lead to final possible solutions. Team members chose the responsibilities and roles of each individual to resolve these issues by creating an issue resolution ladder. Towards the end of the first construction phase of the project, the contractor looked ahead at work that needed to be accomplished at a next stage. Before starting the original second phase (split traffic phase), the contractor asked if he could change the phasing sequence to allow completion of the project at an earlier date. The contractor then submitted a value engineering proposal to the client and designer and was reviewed by the partnering team. During this period of time, the partnering team met on a weekly basis, or more often as needed, to clarify all issues pertaining to combining the second and third phases. The value engineering proposal was approved and the project was completed before the original completion date.

Partnering outcomes

By combining the last two phases, the traffic flow was significantly improved. It was better than the original traffic numbers before the start of project. As a lesson learned, the client subsequently included the evaluation of temporary sidewalks in all bridge projects as a method for reducing the number of phases. The project was completed ahead of schedule and within budget. The elimination of the initial spilt phase increased driver and worker safety during the reconstruction of this interchange. The contractor's work became more efficient after combining the two phases because the contractor was given more working area.

Lessons learned

This project shows that the combination of partnering approach and value engineering management can greatly improve the efficiency of working performance amongst the project team members.

Case study 2

Interstate-95 (I-95) bridge over Brandywine Creek, Wilmington, Delaware – A case study of an infrastructure project in the USA (extracted from the website of AASHTO: http://cms.transportation.org/?siteid=38&pageid=377) (permission has been obtained for both print and online use from the AASHTO).

Background

This bridge project spans 1,875 feet in total, carrying I-95 over the Brandywine Creek in Wilmington, Delaware. In April 2003, a bird-watcher in the park under the bridge noticed a crack in the fascia girder, one of its 12 main girders. An urgent inspection showed that an 8 ft deep fascia girder of the northbound bridge exhibited a significant fracture in the bridge's 245 feet main span. The bottom tension flange was wholly severed and the crack extended more than 80% of the way up the web of the girder.

Major challenges

A number of major challenges were encountered in this project. First, because the stretch of I-95 carried approximately 100,000 vehicles daily, traffic needed to be maintained. This meant that the beam needed to be secured immediately to protect the travelling public and to keep the beam from completely fracturing. The Delaware Department of Transportation (DelDOT) already had a continuous contract to repair the bridge parapets and median barriers and other architectural improvements. The crack was original at a butt-welded splice in a longitudinal stiffener on the fascia girder. Longitudinal stiffeners were only installed on the exterior face of the fascia girders for aesthetic purposes. The project team would ensure that a similar failure would not occur at another location on the bridge. There were a number of options for the permanent repairs. The first alternative was to replace the girder. The second alternative was to lift the cracked girder back to its pre-crack position and splice the girder at that position. The selected permanent repair option was to jack the fractured girder and repair it in place. The next decision was how to jack the girder up into its initial position so it could be repaired. Some methods for doing this were considered. The first method was bridging over the crack from on top of the bridge and lifting the girder. The second one was to construct transverse framing on both sides of the cracked girder and jacking from the framing system. The third option was to jack the beam from temporary towers built in the Brandywine Creek over 80 feet below. The third option was finally selected because it was the most feasible, straightforward and safest choice.

Partnering principles

Representatives from DelDOT were on site within an hour of the call from the bird-watcher. The outside lane was closed to traffic and engineers from DelDOT and the consulting firm immediately inspected the bridge. The University of Delaware joined the project team and installed some strain gauges to monitor the stresses because of live load in the exterior and the first interior girder. It also conducted some diagnostic load tests on both the northbound and southbound bridges. Within two days of the discovery of the crack, a temporary splice plate had been bolted to the tension flange of the fractured girder to recover full capacity to the girder and to prevent the crack from propagating further. All similar butt-welded connections were completely inspected using ultrasonic testing for similar cracking and all the critical sections were retrofitted to prevent additional failures. As the inspectors were locating additional cracks, ironworkers followed behind the inspection crew to complete the retrofit work. All project team members worked cooperatively in the final design of the permanent repair splice so as to minimize the time and cost for design and construction. The University of Delaware worked with the team to set up and carry out a field-testing programme to monitor the permanent repair process and to ensure that the permanent repair was effective in returning the bridge to its prior state. A permit for emergency construction work in the stream was promptly obtained due to the speedy response from Delaware's Department of Natural Resources and Environmental Control, as well as the US Army Corps of Engineers.

Partnering outcomes

Once preparation for jacking was complete, the girder was jacked upward approximately two inches to its pre-crack elevation. As predicted, the crack had closed upon reaching its pre-crack elevation. Once the girder was jacked to its final jacking elevation, the permanent repair splices could be installed. Since the crack did not extend into the top flange, a top flange splice plate was not required. However, one was installed as a catch in the highly improbable case that the crack would continue to spread upward at a later date. A quality control programme was used and monitored to ensure the proper installation of the spice plates. The team completed the permanent splice on time.

Lessons learned

This project shows that partnering approach can assist in enhancing the working relationship amongst the project team members, thus helping to alleviate the challenges and problems encountered in this project.

Case study 3

North east streetscape project – A case study of an infrastructure project in the USA (extracted from the website of the AASHTO: http://cms.transportation. org/?siteid=38&pageid=377) (permission has been obtained for both print and on-line use from the AASHTO).

Background

This Maryland State Highway Administration (MSHA) streetscape project was built through a municipality which has construction of new drainage, curb and sidewalk, bridge work and milling and paving the existing roadway. The project is very sensitive to the needs of local business groups.

Major challenges

A major challenge was to provide a safe and quality project for the travelling public and pedestrians, as well as the residents and businesses of the town. Milling and paving had to be done to the existing roadway along with the construction of parking bump-outs. Because roadway widths were too narrow to accommodate traffic, a detour had to be set up to move traffic through the area. In addition, a concern was that businesses were impacted by a lack of parking, access to front of businesses and overall loss of business revenue.

Initially the contractor was to perform this work during the day. This would permit a clean, quality job during milling and paving and allow for this detailed work to be done under natural light conditions. Great opposition came from the business groups. Before the project began, it was also stated at a town informational meeting that the existing roadway would never be shut down and access to stores would always be available. Unfortunately, the narrow width of the road and opposition caused this issue with the work in question. The question was whether or not this work could be done at night during off-peak hours. This would lessen the impact on the businesses and traffic would be considerably less. On the downside, milling and paving quality would be sacrificed when this work was done at night. The cost was increased and problems with the concrete and asphalt plants running at night were more costly.

Partnering practices

This project fully participated in the partnering programme and practices. A kick-off partnering workshop was held to share information about the partnering concept. All team members allied with the project were invited. Monthly partnering meetings were held to review issues and concerns for the project. During the meetings, all participants filled out the partnering rating forms to evaluate how well the team functioned. This aided in fine-tuning the programme and allowed for a closer examination of issues, concerns and the overall partnering process. Special meetings were also held with individuals who had special concerns or needs. Different town and informational meetings about the project were held to provide information about upcoming construction. To further keep the travelling public and community informed about the work done and proposed for the near future, flyers, handouts, monthly newsletters and newspaper announcements were used.

Partnering outcomes

The aforesaid issue was remedied by having an informational meeting and getting input and concerns from businesses, town official and residents. It was determined to perform this work at night to lessen the impact on the business community and lessen traffic problems that would definitely have occurred because of the high volume of traffic coming through the construction zone. Through the partnering process during the construction of this project, many obstacles, problems and issues were either avoided, or their impact lessened. Modifications to the contract were made to take account of the concerns of all team members in this project.

Lessons learned

This project shows that appropriate use of different partnering tools, such as partnering workshops, regular partnering meetings and special partnering *ad hoc* meetings, can assist in developing a strong and continuous partnering spirit, thus helping to enhance the working relationship amongst the project team members. Ultimately, it could alleviate the challenges and problems encountered in this project.

Case study 4

New Brisbane International terminal building project, Australia – A case study of a large-scale infrastructure project.

Background

The new Brisbane International Terminal Complex was a large-scale infrastructure project and it comprised a four-level steel-framed structure for the concurrent processing of arriving and departing passengers at a rate of 1,200 passengers per hour, apron facilities to accommodate eleven aircraft including eight aerobridge bays, parking for 2,000 vehicles and landscaping of the remaining 55 hectares with plant varieties identified with south east Queensland. The project began in October 1993 and was completed in October 2005. The contract sum was AU$240 million and there were about 600 workers involved in this project, of which 10 came from the clients, 50 from project management/contractors, 100 from consultants and 400 from subcontractors. The key participants included the client (Federal Airports Corporation Brisbane), the head contractor (Civil and Civic Pty Ltd) and the key consultant (Bligh Voller).

Partnering approach and process

Under the traditional form of contract, a client employs a consultant to prepare full documentation. Afterwards, the consultant drafts and calls a tender. The most suitable contractor will then be selected. Under such circumstances, adversarial working relationships between the client, consultant and main contractor is easily induced. Unlike this traditional form of contract, the main contractor in this project was directly employed by the client at first (the project manager came from the main contractor). Then, the main contractor employed the most suitable consultant and subcontractors. This arrangement was always implemented by Federal Airports Corporation Brisbane and it was called relationship and partnering spirit.

Summary results of the end-of-project interview

Major reasons for adopting a partnering approach

The interviewee held that the major reason to adopt a partnering approach in this project was to achieve mutual objectives between the client and the main contractor. One of the most important mutual objectives was to achieve a 'successful' project. The term 'successful' here means value for money and satisfying customers (referring to passengers and airlines).

Perceived major benefits and critical success factors for adopting partnering

To implement partnering, the long-term business relationship was developed amongst the project team members (afterwards, the client won a mega project named Sydney International Terminal Extension in 2000) and the client got profitable returns with good functionality. The most CSF was client satisfaction. In fact, the interviewee stated that it was a means to achieve good cost control with high quality of functionality.

Relationships in partnering

The working relationship between parties was very good. In fact, the key factor that contributed to the high levels of success on this project was the spirit of cooperation that developed between all project team members. Since there were very clear common team goals amongst the project team members, relationships developed between all parties that facilitated the achievement of these goals.

Major difficulties in partnering implementation

The most serious difficulty was with technical complexity. To illustrate the complexity, three primary areas should be discussed, including: (1) project needs; (2) terminal facilities; and (3) construction challenges. The requirements in terms of functional and operational needs of an international airport terminal were many and varied. They comprised a range of building uses that overlapped and required comprehensive coordination and integration to ensure that the terminal's primary function was met (i.e. the efficient transfer of passengers from land transport to air transport whilst facilitating all the processes required for international travel but especially assisting the on-time arrival and departure of aircraft). On the other hand, the physical facilities that met the needs of the terminal's function and operation had high levels of technical complexity. The unique procedures and techniques developed to construct the terminal arose from a range of challenges that involved complex inputs. Although technical complexity was a serious problem encountered in this project, it was resolved by efficient construction coordination between the project team members.

Effect of construction culture on partnering practice and performance

The interviewee perceived that the construction culture in Australia was very wide and varied, but might be summarized as a short sentence: 'people were willing to work together harmoniously in order to get good and successful outcomes'. This culture helped partnering practice in this project because people were easy-going and they were willing to work cooperatively and harmoniously so that adversarial working relationships caused by traditional forms of contract did not happen. The performance of this project was also very good under this culture because team members were willing to work cooperatively, thus optimizing the use of resources.

Effect of organizational culture on partnering practice and performance

The interviewee stated that the organizational culture in his company was that colleagues of all levels were willing to work as one team. He pointed out that the organizational culture in his company was highly recognized in the Australian construction industry. This culture was very supportive to partnering practice in this project because it matched partnering spirit well (based on mutual trust, commitment, innovation and integrity). The culture was also favourable to partnering performance in this project because it added value through collaborative approach.

Lessons learned

The lessons learned from this successful partnering project could assist the practitioners to mitigate the detriments brought about by possible difficulties, and reap the greatest benefits from adopting a partnering approach. This partnering project proved that partnering could lead to improved working relationships amongst project stakeholders through mutually developed strategies towards win-win outcomes for all stakeholders. But one should be aware of the effect of national and organizational culture when implementing partnering.

Case study 5

Tourism and hospital building project at the Palmerston Campus of the Northern Territory University in Darwin, Australia.

Background

This was a small-scale building project with a contract sum of approximately AU$5 million. The project began in 1992 and was completed in 1993. There were about 20 to 30 participants involved in this project. The key participants included the client (Northern Territory University), the main contractor (John Holland Construction) and the main consultant (Woods Bagot).

Partnering approach and process

This partnering project was implemented with 'formal' approach which refers to the development of a partnering charter (it was signed by all participants) and set-up of an issue resolution mechanism during the initial partnering workshop and holding regular partnering workshops and meetings to keep the partnering spirit going on.

Summary results of the end-of-project interview

Major reasons for adopting a partnering approach

The interviewee stated that the major reason for adopting partnering approach in this project was that the client would like to use it to reduce the traditional adversarial working relationship between parties.

Perceived major benefits and critical success factors for adopting partnering

The perceived major benefits included: (1) better working relationships between all parties; (2) focus on quality outcome, not just on price and cost; and (3) satisfaction of all parties.

The CSF for adopting partnering in this project was that all parties were willing to commit to the partnering charter and maintain it during the whole life of the project.

Relationships in partnering

The working relationship between parties was good in that the client was willing to accept the project team's ideas and suggestions.

Major difficulties in partnering implementation

The major difficulty encountered in this project was the tight budget. However, it was resolved by the client because he was willing to accept the reduction of the scope of the project.

Effect of construction culture on partnering practice and performance

The interviewee pointed out that the construction culture varied from state to state. The relationship between builders and subcontractors was very confrontational in Victoria and Western Australia because the unions were very strong. However, the relationship between subcontractors and builders was much better and more cooperative in states like Queensland and Northern Territory because the unions were not so strong. Since the construction culture was more cooperative in Northern Territory, it supported partnering practice and performance in this project.

Effect of organizational culture on partnering practice and performance

The interviewee viewed that the culture in his organization was characteristic of: (1) a global company structure; (2) a number of global partners; (3) lots of cooperation between offices; and (4) directors were willing to share human resources between offices all over the world.

This organizational culture helped partnering practice and performance in this project because parties had a common interest to share the benefits (they were willing to work together to get a good outcome).

Lessons learned

This case study has indicated that partnering can also be implemented successfully for small-sized projects and they could enjoy great partnering benefits.

Case study 6

Bruce Highway Yandina to Cooroy project, Australia – A case study of a medium-sized infrastructure project.

Background

This was a medium-sized infrastructure project that built a highway based on an existing two lane road. The project began in 1998 and was completed in 2000. The contract sum was approximately AU$70 million and there were about 100 workers involved in this project, of which two were from the client (Department of Main Roads in Queensland), 20-30 from the contractor (Leighton Contractors), five from the superintendent and 60-70 from subcontractors.

Partnering approach and process

The partnering approach in this project was initiated by the Department of Main Roads who first put 'partnering' into the tender documents. After that, the Department of Main Roads paid some money to staff of Leighton Contractors to attend partnering workshop and follow-up partnering meetings. Unlike the usual partnering approach, this project has applied so-called 'extended partnering' approach. Main Roads Project Delivery System (2005) stated that extended partnering is a formal process used to facilitate greater team participation and communication outside of the contractual process. It is used to develop a cooperative approach between all parties to the contract to achieve best project outcomes. It does this by achieving some initial impetus towards positive working relationships and shared goals at the start of the project, providing appropriate problem-solving and team-building skills to the parties to achieve and maintain these, and providing an agreed structure, which assists in maintaining and improving these relationships. Figure 8.2 shows the typical process of the extended partnering.

Summary results of the end-of-project interview

Major reasons for adopting a partnering approach

The interviewee stated that the major reason for adopting a partnering approach was due to a previous successful project, named 'Pacific Motorway' which adopted a partnering approach.

Perceived major benefits and critical success factors for adopting partnering

The interviewee reckoned that the major benefits for adopting partnering were: (1) very good working relationships with other parties; (2) every problem was solved in the partnering meetings; (3) issues were resolved quickly; (4) no outstanding claims or issues; and (5) project was completed on time and within budget.

The most CSF for adopting partnering approach in this project was support from top management. The deputy director general had shown great support for this project in that he attended almost all partnering meetings. In return, a person with similar position from Leighton Contractors also attended nearly all partnering meetings. Therefore, strong partnering spirit was built.

Relationships in partnering

The working relationship between parties was very good in that industrial practitioners were prepared to share problems and solve them together.

Major difficulties in partnering implementation

Project team members had encountered some technical difficulties and some unsuitable materials were bought, but this problem was resolved smoothly in the partnering meetings.

Effect of construction culture on partnering practice and performance

The interviewee perceived that the construction culture in Australia was changing from adversarial to more cooperative due to the occurrence of partnering, alliancing and RC in recent years. Industry practitioners might gradually come to believe that cooperation is much better than competition.

Since industry practitioners had generally gone through adversarial working relationships with other parties and gradually experienced the benefits of adopting partnering approach, they (both project participants from the Department of Main Roads and Leighton Contractors) believed that partnering was better than the traditional way to procure this project.

As the Australian culture was changing from an adversarial working relationship to be more cooperative, practitioners experienced more success. Such a business environment helped this project with better performance (as the interviewee mentioned: success leads to success and then leads to more and greater success.)

Effect of organizational culture on partnering practice and performance

The interviewee stated that the organizational culture in his company was changing to be more cooperative (due to the fact that contractors were willing to be more cooperative). This culture matched the partnering spirit very well so that this project ended up with good performance in terms of innovations, working relationships, time and cost. It was also very supportive because senior management staff were more willing to commit and they were more adaptive to new ideas. They did not treat other parties as the enemy. Therefore, it was easy for them to bring ideas forward.

Lessons learned

This case study has provided valuable insights that the 'extended' partnering approach as applied in the infrastructure sector in Australia could generate stronger teamwork and cooperation between contracting parties, thus resulting in better overall project performance. It is concluded that extended partnering may be a preferred model to facilitate greater team participation and communication, and more cooperative working environment outside of the traditional contractual process.

Case study 7

City West Link project, Australia – A case study of an infrastructure project.

Background

This was an infrastructure project (a road project), which was built along a medium-density residential area in New South Wales. It was a four-lane highway with bridge structures and the highway was 4 km long. The project began in 1997 and was completed in 2000, with a contract sum of around AU$190 million. There were about 250 workers involved in this project, of which seven were from client, 20 from consultants, 80 from the main contractor and 140 from subcontractors. The client was Roads and Traffic Authority (RTA) in New South Wales and the main contractor was John Holland Construction.

Partnering approach and process

The interviewee mentioned that RTA and John Holland did some business previously. The client was responsible for its own design (or let consultants do it), and the contractors were responsible for construction works only. However, it was the first time (for a number of years) for this project to use design-and-construct contract so that both the client and the contractor had more cooperation, thus leading to better working relationship and collaboration.

Summary results of the end-of-project interview

Major reasons for adopting a partnering approach

The major reason for adopting a partnering approach in this project was to get the best outcomes in terms of time, cost, quality, safety and innovation.

Perceived major benefits and critical success factors for adopting partnering

The interviewee perceived that to adopt partnering, the major benefit for the main contractor was to be able to get more exposure to the client so that the client would be comfortable in giving further work. The client received a very good project, e.g. within time and budget and of high quality.

The CSFs for adopting partnering in this project were to get community engagement (fewer complaints for the residents living in the vicinity) and to have a good mechanism for resolution of conflict amongst different parties.

Relationships in partnering

The working relationship between parties was very good in that when there were divided opinions amongst industrial practitioners, they were willing to respect to the others.

Major difficulties in partnering implementation

The interviewee was not sure what the major difficulties had encountered in this project because he was not involved at that level.

Effect of construction culture on partnering practice and performance

The interviewee perceived that the construction culture varied from state to state. He believed that it was very confrontational in some states, such as Victoria and Western Australia but was more cooperative in some states, such as Queensland. The construction culture mentioned above did not affect partnering practice and performance in this project much.

Effect of organizational culture on partnering practice and performance

The interviewee pointed out that it was quite difficult for him to comment on the organizational culture in his company because he was only there for one year and did not have much exposure to remainder of the company. But he stated that project participants were quite cooperative with outcome-focused direction and this project culture helped set a good network amongst different parties. On the other hand, this project culture was very positive for achieving good partnering performance because project participants were very cooperative and showed high level of respect with each other much. In addition, they formed an integrated team and more innovations were created.

Lessons learned

Like other successful partnering projects adopted in Australia, this partnering project reconfirmed that a partnering approach could help to improve working relationships among project stakeholders through mutually developed strategies towards win-win outcomes for all stakeholders.

Case study 8

Ballarat Bypass stages 3 and 4 project, Australia – A case study of a small-sized infrastructure project.

Background

This was a small-scale infrastructure project (a government road project) and it was procured by design-and-build approach. The project began in 1994 and was completed in 1998, with a contract sum of approximately AU$9 million for stage 3 and AU$18 million for stage 4. There were around 50 participants in this project, including 10 from the client, 15 from the main contractor, and 25 from subcontractors. The client for stages 3 and 4 was Vic Road, the main contractor for stage 3 was Cook Construction Pty Ltd and the main consultant for stage 3 was Ove Arup Ltd. The head contractor for stage 4 was John Holland and the main consultant was Maunsell Ltd.

Partnering approach and process

At first, a partnering charter based on mutual trust and commitment amongst parties was developed through an initial partnering workshop. There were common goals, open commitment and a dispute resolution mechanism developed through the partnering charter. After that, friendship was maintained through follow-up regular partnering meetings.

Summary of the end-of-project interview

Major reasons for adopting a partnering approach

The major reasons for adopting a partnering approach in this project were that Vic Road would like to improve the working relationship with the project team (to reduce claims) and the contractors would like to get higher returns.

Perceived major benefits and critical success factors for adopting partnering

The perceived major benefits for implementing partnering included: (1) improving communication; (2) building up mutual trust; and (3) problems were resolved quickly and smoothly. The critical success factor for adopting partnering in this project was mutual trust and respect between parties.

Relationships in partnering

The working relationship between the client and the project team was good because most of the project team members worked cooperatively.

Major difficulties in partnering implementation

A major difficulty was to maintain harmonious working relationships when conflicts arose. However, the problems were resolved by open communication during partnering meetings.

Effect of construction culture on partnering practice and performance

The interviewee opined that the construction culture in Australia was very adversarial under the traditional form of contract. In fact, this national culture was not good for this project adopting partnering approach.

Effect of organizational culture on partnering practice and performance

The interviewee viewed that many people in Vic Road worked under very high pressure. There was serious shortage of skilled people (many old and young people and there was a gap between them) and many projects were managed by inexperienced people and this organizational culture made this partnering project more difficult to succeed.

Lessons learned

This case study has reconfirmed that partnering can also be implemented successfully for small-sized projects and they could reap full partnering benefits.

Case study 9

Ipswich Hospital redevelopment project, Australia – A case study of a medium-sized hospital project.

Background

This was a medium-sized hospital project, with a contract sum of around AU$85 million. The project began in 1999 and was completed in 2001. There were about 200 participants involved in this project, of which six came from the client, 20 from the main contractor, 30 from consultants and 140 from subcontractors. The client was Queensland Department of Health, the main contractor was Walter Construction Group (a German company), and a main consultant was Woods Bagot Consultant Ltd.

Partnering approach and process

At first, the main contractor approached the client and persuaded him to adopt a partnering approach to procure this project and the client agreed. After that, partnering workshops were facilitated and a partnering charter was signed by all project stakeholders and an issue resolution mechanism was set up. Follow-up partnering meetings were held regularly to monitor the project performance.

Summary results of the end-of-project interview

Major reasons for adopting a partnering approach

The major reason for adopting a partnering approach in this project was to overcome confrontational working relationship amongst different stakeholders and look for better ways to build up better working relationships between parties.

Perceived major benefits and critical success factors for adopting partnering

There were better relationships with the other organizations and this project was enjoyable to work on. The critical success factor for adopting partnering approach was the high level of communication, thus inducing high level of mutual trust and good coordination.

Relationships in partnering

The working relationship between the project team members was very good.

Major difficulties in partnering implementation

The interviewee perceived that there were no major difficulties encountered in this project when partnering was implemented.

Effect of construction culture on partnering practice and performance

The interviewee stated that the effect of construction culture in Australia on partnering practice and performance depended on the project and method of delivery and the construction culture did not affect partnering practice and performance much in this project.

Lessons learned

Similar to other successful partnering projects adopted in Australia, this partnering project reconfirmed that partnering approach could induce to more harmonious working relationships among project stakeholders through mutually developed strategies towards win-win outcomes for all stakeholders.

Case study 10

Chater House, Hong Kong – A case study of a prestigious office development project (Source from Chan et al., 2004b) (permission has been obtained for both print and online use from Construction Industry Institute, Hong Kong).

Background

The project was composed of a demolition of an existing building (Swire House) and a construction of a 29-storey international Grade A office development in Hong Kong's central business district, which comprised a 3-storey basement, a 3-storey podium and a 23-storey tower (Figure 11.1). The overall gross floor area was 74,000m². The contract sum was approximately HK$1.2 billion (approximately US$154 million), with original contract duration of 641 days. The project was procured by a negotiated guaranteed maximum price (GMP) contract. The mechanism of the GMP contract requires that the client, consultants and the main contractor work as a team to determine construction method, programme, pricing, detailed breakdown of direct works and consent to preliminaries and conditions of contract (Chan et al., 2004b). This entailed the main contractor to release all his back-up data to team members. The exchange of this information required a high level of trust amongst the team, especially the main contractor.

The major project stakeholders included the following organizations:

Client	Hongkong Land Ltd (HKL)
Main Contractor	Gammon Skanska Ltd
Design Architect	Kohn Pedersen Fox Associates
Project Architect	Aedas LPT Ltd
Structural Engineer	Ove Arup & Partners (HK) Ltd
Electrical & Mechanical Consultant	WSP Hong Kong Ltd
Quantity Surveyor	WT Partnership (HK) Ltd

Partnering approach and process

Figure 11.2 shows the partnering approach and process of Chater House in which there were a total of three partnering workshops, encompassing one initial workshop, one interim workshop and one final workshop. The initial workshop, lasting for 1.5 days was held at four per cent post-contract period with 42 participants mainly coming from the client's and the main contractor's companies. It is of interest to note that unlike the United States and other advanced countries where the first or inaugural partnering

Figure 11.1 Outlook of Chater House

workshop is usually held after a contract award but generally before any contract work is initiated, the initial partnering workshop in this project is held after the contract work started. Eight activities were undertaken with the guidance of an external facilitator, including: (1) discussions of goals and objectives; (2) strengths of partners and obstacles to success; (3) brainstorming on how to overcome obstacles; (4) soliciting ideas to make partnering work; (5) development of the partnering charter and the action plan; (6) participation in a game; (7) setup of the problem-resolution process; and (8) nomination of partnering champions.

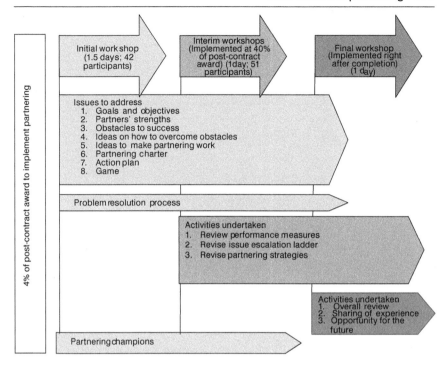

Figure 11.2 The partnering approach and process of Chater House (adapted from Latham's (1994) Report; source: Chan et al., 2004b)

Results on key performance indicators

Chan and Chan (2004) developed a framework to measure the success of construction projects in which a set of KPIs were measured both objectively and subjectively. The KPIs include time, speed, time variation, construction cost, incident rate and number of environmental complaints.

Table 11.1 shows the KPIs for Chater House. It is noted that the time variation was ahead of schedule by 0.63% and the construction cost was within budget. The incident rate of this project was 17/1000, which was much lower than the industry average of 85.2/1000 based on the statistics released by the Hong Kong Labour Department in 2002. All these KPIs suggested that this was a successful partnering project.

Lessons learned

The lessons learned from this successful partnering project could help the practitioners to minimize the detriments brought about by possible difficulties, and maximize the benefits gained from implementing partnering concept. Such an improved understanding could generate essential strategies

Table 11.1 Key performance indicators (KPIs) for Chater House Project (reproduced from Chan et al., 2004b) (with permission for both print and online use from Construction Industry Institute – Hong Kong)

KPIs	Chater House
1. Construction time	635 calendar days
2. Speed of construction	84.72m²/day
3. Time variation	-0.63%
4. Construction cost	Within budget
5. Incident rate	17/1000
6. No. of complaints received caused by environmental issues	Nil

to eradicate the root causes of poor project performance and ineffective communication (Chan et al., 2002a). The partnering approach could lead to improved working relationships amongst project stakeholders through mutually developed strategies towards win-win outcomes for all stakeholders. The partnering process empowers the project personnel to accept responsibility and to do their jobs by delegating decision making and problem solving to the lowest possible level of authority (Dunston and Reed, 2000). These findings reinforce the recommendations of the Latham Report (1994) published in the UK, which forms the basic guidelines and skeleton for implementing successful partnering in future projects. Emphasis should be put on the early implementation of partnering process and a structured approach, the careful design of partnering activities, regular monitoring of partnering process, selection of qualified facilitators for partnering workshops, and appointment and true empowerment of the partnering champions (Chan et al, 2004b).

It is concluded that project partnering, when implemented properly, can generate a workable model for people to communicate more effectively and efficiently thus eliminating unnecessary misunderstanding and possible conflicts. Therefore, partnering should be adopted across a wider spectrum of the construction industry to reap sustainable benefits and achieve construction excellence.

Case study 11

One Peking Road project, Hong Kong – A case study of a completed deluxe commercial building (Source from Chan et al., 2004e) (permission has been obtained for both print and online use from Construction Industry Institute, Hong Kong).

Figure 11.3 Outlook of One Peking Road

The major project stakeholders included the following organizations:

Client: Glorious	Sun Holdings Ltd
Main Contractor	Gammon Skanska Ltd
Project Manager	DTZ Debenham Tie Leung
Project Architect & Structural Engineer	WMKY Ltd
Design Architect	Rocco Design Ltd
Electrical & Mechanical Engineer	J Roger Preston Ltd
Quantity Surveyor	Levett & Bailey
Curtain Wall Consultant	Heitmann & Associates Hyder Consulting Ltd
Curtain Wall Contractor	Permasteelisa Hong Kong Ltd
Lighting Consultant	TinoKwan Lighting Consultant

Background

Glorious Sun Holdings Ltd, the developer of One Peking Road Project, is a diversified multinational enterprise, which was set up in Hong Kong in 1974 with core businesses on textiles and apparel, property development, financial investment, and interior decoration and renovation services.

One Peking Road is a challenging building project. Glorious Sun Holdings Ltd appointed Gammon Skanska Ltd and Permasteelisa Hong Kong Ltd as the main contractor and curtain wall contractor respectively. The completed deluxe commercial building provides a 30-storey world-class office space and retail development with a 3-level shopping arcade and a 4-level basement. The gross floor area is 33,401m², with a contract sum of HK$550 million and a superstructure construction period of 19 months. Its characteristics include: (1) fantastic uninterrupted views across Victoria Harbour to Hong Kong's central business district; (2) exceptionally accessible and high profile location, as evidenced by early occupancy commitment from retailers such as Escada, Christian Dior, Cartier, Fendi and Zegna; (3) close to shops and prestigious hotels; (4) very attractive leasing incentives; and (5) cutting-edge building specifications including 2.8m clear headroom for ceiling, 160mm raised floor and 470mm of space under the beams for air ducts.

At the very beginning of the project, Glorious Sun and Gammon had already decided to adopt the partnering approach, which aimed to deliver the best project performance to all stakeholders in terms of time, cost, quality, safety and environmental objectives, as well as client satisfaction. By doing so, they greatly demonstrated the significant ingredients of the partnering spirit, including commitment from senior management, treating each other on equal basis, mutual trust and respect, setting common goals and implementing a win-win philosophy. As a result, in addition to securing very high-quality commercial premises, it was completed on time and within budget. The daily travel rate was approximately 11,000 persons, and all shops were successfully leased within six months. The financial returns could be achieved in a short time and were higher than what the client expected. The industrial practitioners from professional bodies admired the project very much, and project team members felt a sense of pride and encouragement. The project had even unexpectedly won five outstanding awards across the whole construction industry, including the Quality Building Award 2004 under the non-residential building category amidst very fierce competition with its counterparts.

The partnering approach

The One Peking Road project adopted an unstructured partnering approach with strong partnering spirit. Most of the colleagues were highly respectful and trustful of each other in that no matter how high or low their professional affiliations and levels, they could voice their views during the bi-weekly regular meetings. Many problems were thus resolved swiftly because many colleagues at various levels were willing to attend the meetings and took the responsibility to solving the problems in which they were involved. Moreover, the project was undertaken based on people-focused approach instead of system-focused approach. One of the client representatives believed that the Christian spirit was vital to the success of the project because everyone did his or her best work. This partnering spirit underpinned the success of unstructured partnering approach adopted in the One Peking Road project.

Critical success factors for unstructured partnering

Six CSFs for One Peking Road project were identified to be significant during the group interview meeting based on the perceptions of the client representatives, main contractor representatives, consultant representatives and subcontractor representatives. The six CSFs encompassed: (1) mutual trust; (2) respect; (3) equity; (4) people-focused approach; (5) good leadership; and (6) early implementation of partnering process (Figure 11.4).

Lessons learned

This case study has generated abundant insights into enhancing overall project performance due to the successful implementation of a partnering project with an unstructured approach. It is clear from the case study of One Peking Road project that structured approach is not 'a must' for a successful partnering venture, but strong partnering spirit underpins the success of a partnering project no matter whether it adopts structured approach or unstructured one. Structured approach with strong partnering spirit, however, makes a partnering project more successful because it can provide extensive benefits to the industry across different sectors. It must, however, be emphasized that partnering is not a panacea. Instead, it only

Figure 11.4 Critical success factors for the One Peking Road project

provides an avenue where people can communicate better thus eradicating unnecessary misunderstanding. Nevertheless, it is still highly recommended that partnering be adopted across a wider spectrum of the construction industry to reap sustainable benefits for construction excellence during the twenty-first century.

Case study 12

1063 King's Road, Hong Kong – A case study of a quality commercial building (Source from Chan et al., 2004b) (permission has been obtained for both print and online use from Construction Industry Institute, Hong Kong).

Figure 11.5 Outlook of 1063 King's Road

Background

The scope of work comprised the construction of a 30-storey quality commercial office. The building had limited ground floor retail space is situated in the district of Quarry Bay. The original contract sum was HK$375 million (equivalent to about US$48.4 million), with original contract duration of 636 calendar days (from November 1997 to August 1999). The participants included:

Client	Hongkong Land Ltd (HKL)
Main contractor	Gammon Skanska Ltd
Architect	Wong Tung & Partners Ltd
Structural engineer	Maunsell Consultants Asia Ltd
Façade consultant	Meinhardt Façade Technology (HK) Ltd
Mechanical & Electrical consultant	Meinhardt E&M Ltd
Acoustic consultant	Shen Milsom & Wilke Ltd
Quantity surveyor	WT Partnership (HK) Ltd

The project was procured by a negotiated GMP contract. The project site is within the Mass Transit Railway Corporation Limited (MTRCL) protection boundary. The bored-pile foundation was carefully arranged to avoid imposing any undue surcharge to the tunnel. The site maximized the permissible site coverage.

Unstructured partnering approach

The acquisition transaction for the site was concluded in September, 1996 and the lot was handed over to Hongkong Land Ltd (HKL) at the end of December, 1996. The consultant team was required to develop a demolition plan and the approval was obtained from Buildings Department (BD) to allow the demolition contractor to start work upon receipt of land at the end of December, 1996. This was achieved through good teamwork among major project consultants with close liaison with BD. It was worth noting that Hong Kong had a number of demolition accidents and all demolition proposals were being reviewed carefully by the BD at that moment.

It was obvious to HKL that the traditional contract arrangement had certain drawbacks. Therefore, the client was determined to develop a win-win partnering approach to manage the project. A value engineering session was conducted with the participation of all relevant parties including the contractor and the client's property management personnel.

The mechanism of the GMP contract was envisaged and required the major stakeholders to work as a team to determine the construction method, the programme, pricing details, preliminaries and conditions of contract.

Table 11.2 Key performance indicators (KPIs) for 1063 King's Road Project (reproduced from Chan et al., 2004b) (with permission for both print and online use from Construction Industry Institute – Hong Kong)

KPIs	1063 King's Road Project
1. Actual construction duration	636 calendar days
2. Speed of construction	42.92 m²/day
3. Time variation (actual duration – contract duration)	0% (i.e. completion on schedule)
4. Cost variation (actual cost – contract sum)	Within budget
5. Injury (accident) rate	12.8/1000 employees
6. No. of environmental complaints received	Nil

This required the main contractor to release all the back-up data to the team members. The exchange of this information required a high level of trust amongst the teams, including the main contractor.

During the selection of nominated subcontractors and domestic subcontractors, the client, the consultants and the main contractor were free to make comments and give advice and recommendations in a spirit of getting the most competent subcontractors for the project to obtain the desired quality building.

A spirit of trust was developed between the main contractor and the project team during the pre-contract stage, which enabled all the team members to work in a win-win partnering spirit. Members became more focused on problem solving rather than shifting responsibility to each other.

The team spirit developed during the design stage was successfully extended to the construction stage. As the contractor team had the opportunity to participate in an early design stage, the personal trust developed amongst the team members could be smoothly transferred to site level and extended throughout the whole contract period. Correspondence was more constructive in getting matters solved instead of being contractual.

Results on key performance indicators

The KPIs used in this project include: (1) construction time; (2) speed of construction; (3) time variation; (4) cost variation; (5) injury (accident) rate; and (6) number of environmental complaints received. Table 11.2 shows that the time variation for 1063 King's Road Project was 0% implying completion on schedule and the cost variation was within budget. The injury (accident) rate of this project was 12.8/1000 employees, which was much lower than the industry average of 85.2/1000 based on the statistics released by the Hong Kong Labour Department in 2002 (Hong Kong Labour Department, 2004). All these KPIs indicated that this was a successful partnering project.

Lessons learned

This case study has reconfirmed that it is unnecessary to adopt a structured approach for a successful partnering venture. In contrast, strong partnering spirit underpins the success of a partnering project no matter whether it adopts structured approach or unstructured one. In addition, it has been found that GMP is contributory to partnering success.

Case study 13

Tuen Mun Area 4C, Hong Kong – A case study of a public sector housing project (Source from Chan et al., 2004b) (permission has been obtained for both print and online use from Construction Industry Institute, Hong Kong).

Background

The project is composed of constructing four high-rise towers of residential blocks intended as subsidized housing in Tuen Mun. Each block has 36 storeys and is built on the landscape podium. In addition, there are four levels of podium car park, hard landscape works on the ground floor and podium floor, access road construction and green area formation works.

The original contract sum at tender award was HK$596,739,000 with original construction duration of 650 calendar days. The key participants included:

Client	Hong Kong Housing Society
Main contractor	Hsin Chong Construction Co. Ltd
Architect	Dennis Lau & Ng Chun Man Architects and Engineers (HK) Ltd
Structural engineer	Chung and Ng Consulting Engineers Ltd
Electrical & Mechanical consultant	Gregory Asia Building Services Consulting Engineers
Quantity surveyor	D.G. Jones & Partners (HK) Ltd

The contract was procured on a traditional approach.

Results on key performance indicators

The KPIs used for this case study include time, speed, time variation, construction cost, injury (accident) rate and number of environmental complaints received. Table 11.3 shows that the time variation was 0% and the construction cost was within budget. The injury (accident) rate of this project was 2/1,000 employees, which was much lower than the industry

Figure 11.6 Outlook of Tuen Mun Area 4C (Kingston Terrace)

average of 85.2/1000 employees based on the statistics released by the Hong Kong Labour Department in 2002.

Summary results of interviews regarding major difficulties in implementing partnering

The client representative stated that the consultants were reluctant to implement partnering because they needed to do more work but the benefits could not be easily realized. The main contractor representative believed that his company had no problems in dealing with the client. However, his organization faced some difficulties in educating its subcontractors about the need to meet quality requirements in the first place. The consultant representative considered that he spent too much time on meetings instead of supervising the project. Also, the architect felt an implicit expectation on him to relax standards. The client and the main contractor had to deal with the grey areas on the contract.

Table 11.3 Key performance indicators (KPIs) for Tuen Mun Area 4C Project (Kingston Terrace) (reproduced from Chan et al., 2004b) (permission has been obtained for both print and online use from Construction Industry Institute – Hong Kong)

KPIs	Tuen Mun Area 4C
1. Actual construction duration	835 calendar days
2. Speed of construction	85.15m²/day
3. Time variation (actual duration – contract duration)	0% (i.e. completion on schedule)
4. Construction cost	Within budget
5. Injury (accident) rate	2/1000 employees
6. No. of environmental complaints received	2

The subcontractor representative viewed that the client faced problems with budget control whereas the main contractor had to meet the schedule. On the other hand, the subcontractors faced with the usual problems and partnering did not have any effect on their performance.

Results on ranking of partnering attributes concerning major difficulties in adopting partnering

Chan et al. (2003a, b) carried out the mean score (MS) method to establish the relative importance for partnering implementation in the Hong Kong construction industry. A similar method was used in this section to indicate the ranking of major difficulties in implementing partnering. The five-point Likert scale (from 1 = Strongly Disagree, 2 = Disagree, 3 = Neutral, 4 = Agree, and 5 = Strongly Agree) was used to calculate the mean scores for the major difficulties. Ten major difficulties were elicited from the results of the previous empirical survey undertaken by Chan et al (2003b) and they formed one part of the current questionnaire to examine the perceptions of project participants towards major difficulties in implementing partnering (Table 11.4). Respondents were asked to rate their degree of agreement against each of the identified major difficulties according to a five-point Likert scale.

Table 11.4 shows that the top three major difficulties are: parties were faced with commercial pressure; little experience with the partnering approach; and dealing with large bureaucratic organizations, followed by risks and rewards were not shared directly; and partnering concepts not fully understood by the participants. It is of interest to note that 90% of the items are higher than or equal to 3.5, which mean that this case study project experienced difficulties in implementing partnering. The reason behind this is perhaps due to the nature of the project: the public sector is known to

Table 11.4 Ranking for the major difficulties of partnering projects for Tuen Mun Area 4C (Kingston Terrace) (reproduced from Chan et al., 2004b) (with permission for both print and online use from Construction Industry Institute – Hong Kong)

Major difficulties in implementing partnering	Mean score
Parties were faced with commercial pressure which compromised the partnering attitude	4.25
The parties had little experience with the partnering approach	4.00
Dealing with large bureaucratic organizations impeding the effectiveness of partnering	3.75
Risks or rewards were not shared directly	3.75
The concept of partnering was not fully understood by the participants	3.75
Uneven levels of commitment were found amongst the project participants	3.50
Conflict arose from misalignment of personal goals with the project goals	3.50
Parties did not have proper training on partnering approach	3.50
Participants were conditioned in a win-lose environment	3.50
The partnering relationship created a strong dependency on other partners	3.25

be less flexible and has more stringent procedures to follow. Emphasis on accountability may reduce flexibility to some extent and it will hinder the successful implementation of partnering concepts.

Lessons learned

The research findings related on this case study are useful in understanding the obstacles in implementing project partnering in the public sector of Hong Kong. It is found that the public sector experienced more obstacles in implementing project partnering due to its organizational nature with less flexibility and more stringent and bureaucratic procedures to follow when variations arise.

Case study 14

Kai Tak Estate redevelopment, phase 2, Hong Kong – A case study of a public sector housing project (Source from Chan et al., 2004b) (permission has been obtained for both print and online use from Construction Industry Institute, Hong Kong).

Background

The project comprised the construction of two residential towers (blocks 4 and 5) intended as subsidized housing at Wong Tai Sin, Kowloon, with 26 storeys and 33 storeys respectively rising from the transfer plates. The top two podium floors contain commercial and government and institutional community (GIC) areas, below which is a two-level basement car-park.

The original contract sum at tender award was approximately HK$300 million (equivalent to about US$38.67 million), with original construction duration of 580 calendar days. The key participants included:

Client	Hong Kong Housing Society
Main contractor	Gammon Construction Ltd
Architect	Wong & Ouyang (HK) Ltd
Structural engineer	Siu Yin Wai & Associates Ltd
Electrical & Mechanical consultant	Hyder Consulting Ltd
Quantity surveyor	Widnell Ltd

The contract was procured on a traditional design-bid-build approach.

Results on key performance indicators

The KPIs for Kai Tak Estate redevelopment phase 2 are indicated in Table 11.5 from which the time variation was 0% and the construction cost was within budget. The injury (accident) rate of this project was 28/1,000 employees, which was much lower than the industry average of 85.2/1,000 based on the statistics released by the Hong Kong Labour Department in 2002.

Table 11.5 Key performance indicators (KPIs) for Kai Tak Estate Redevelopment Phase 2 (reproduced from Chan et al., 2004b) (with permission for both print and online use from Construction Industry Institute – Hong Kong)

KPIs	Kai Tak Estate Redevelopment Phase 2
1. Actual construction duration	702 calendar days
2. Speed of Construction	73.85 m²/day
3. Time Variation (actual duration – contract duration)	0% (i.e. completion on schedule)
4. Construction cost	Within budget
5. Injury (accident) rate	28/1000 employees
6. No. of environmental complaints received	Nil

Figure 11.7 Outlook of Kai Tak Estate Redevelopment Phase 2

Summary results of interviews regarding major difficulties in implementing partnering

The client stated that the traditional culture was different and the participants were unable to understand the concept fully and many misinterpretations created problems. Besides, he opined that the main contractor might face the problem of partnering education. The main contractor observed insufficient trust in this project and doubted the possibility of partnering success by junior level staff. The consultant stated that he observed no serious problem, but he considered that the contractors were affected by tight schedule and budget. The subcontractor regarded that it was difficult to monitor the partnering performance because some participants did not express their real feelings. On the other hand, more workload was created and the participants used more time and resources in the partnering practice.

Table 11.6 Ranking for the major difficulties of partnering projects for Kai Tak Estate Redevelopment Phase 2 (reproduced from Chan et al., 2004b) (with permission for both print and online use from Construction Industry Institute – Hong Kong)

Major difficulties in implementing partnering	Mean score
Dealing with large bureaucratic organizations impeding the effectiveness of partnering	4.50
Parties were faced with commercial pressure which compromised the partnering attitude	4.00
The parties had little experience with the partnering approach	4.00
Uneven levels of commitment were found amongst the project participants	4.00
Parties did not have proper training on partnering approach	4.00
Risks or rewards were not shared directly	3.75
The concept of partnering was not fully understood by the participants	3.75
The partnering relationship created a strong dependency on other partners	3.75
Conflict arose from misalignment of personal goals with the project goals	3.50
Participants were conditioned in a win-lose environment	3.50

Results on ranking of major difficulties in adopting partnering

The mean score (MS) method (Chan et al., 2003a, b) mentioned previously was used in this section to indicate the ranking of major difficulties in implementing partnering in this public housing project. The five-point Likert scale (from 1 = Strongly Disagree, 2 = Disagree, 3 = Neutral, 4 = Agree, and 5 = Strongly Agree) was used to calculate the mean scores for the major difficulties.

Table 11.6 shows that the top three major difficulties are: (1) dealing with large bureaucratic organizations; (2) parties were faced with commercial pressure; and (3) little experience with the partnering approach, followed by uneven levels of commitment and lack of proper training on the partnering approach. It is of interest to note that all the items are higher than or equal to 3.5, which mean that this case study project experienced much difficulty in implementing partnering.

Lessons learned

This case study has reconfirmed that the major difficulties encountered in the public sector projects of Hong Kong are probably attributed to the nature of the project; the public sector is known to be less flexible and has more stringent procedures to be followed whenever variations occur. Emphasis on accountability may reduce flexibility to a certain extent and it will hinder the successful implementation of partnering concepts in this sector of projects. The public sector client organizations need to overcome all these identified potential difficulties to accrue the maximum benefits derived from a partnering philosophy.

Case study 15

MTRCL Contract 601 (Hang Hau Station and Tunnels), Hong Kong – A case study of an infrastructure sector project (Source from Chan et al., 2004b) (permission has been obtained for both print and online use from Construction Industry Institute, Hong Kong).

Background

The Mass Transit Railway Corporation Limited (MTRCL) Tseung Kwan O Extension (TKE) Contract 601 (Hang Hau Station and Tunnels) is one of the 13 civil contracts and involved the construction of Hang Hau Station, one property podium structure (including a public transport interchange), two ancillary buildings, as well as cut-and-cover tunnels. Figure 11.13 shows the outlook of Hang Hau Station and Tunnels. The contract sum at tender was approximately HK$1.3 billion, with original duration of 1,246 calendar days. The key participants included:

Client	Mass Transit Railway Corporation Limited (MTRCL)
Main Contractor	Dragages et Travaux Publics (HK) Ltd (DTP
Consultants	Parsons Brincherhoff (Asia) Ltd and Scott Wilson (HK) Ltd

The project adopted the lump sum fixed price contract.

Partnering approach and process

Figure 11.9 shows the partnering approach and process of MTRCL TKE Contract 601 (Hang Hau Station and Tunnels) in which there were a total of seven partnering workshops, which included one initial workshop, five interim workshops and one final review. The two-day initial workshop was facilitated by an external facilitator at 6% of post-contract award period with 24 participants mainly coming from the client and the main contractor. Fourteen activities were undertaken with the guidance of an external facilitator.

Figure 11.8 Outlook of MTRCL Tseung Kwan O Railway Extension (TKE). Contract 601 (Hang Hau Station and Tunnels).

Results on key performance indicators

Table 11.7 shows the KPIs for the MTRCL TKE Contract 601 (Hang Hau Station and Tunnels). It is of interest to note that the time variation was ahead of schedule by 5.62% and the construction cost was within budget. The accident rate of this project was 30.54/1000 employees, which was much lower than the industry average of 85.2/1000 based on the statistics released by the Labour Department of the Hong Kong Special Administrative Region (HKSAR) Government in 2002. All these KPIs reflected that the MTRCL TKE Contract 601 was a successful partnering project.

Lessons learned

This case study has provided valuable lessons to help industry practitioners to minimize the drawbacks brought about by potential difficulties, and maximize the benefits gained from implementing partnering principle. Although partnering is not a panacea, it provides an avenue where project participants can communicate better thus eliminating unnecessary misunderstanding. For infrastructure projects in particular, where solution of method-related

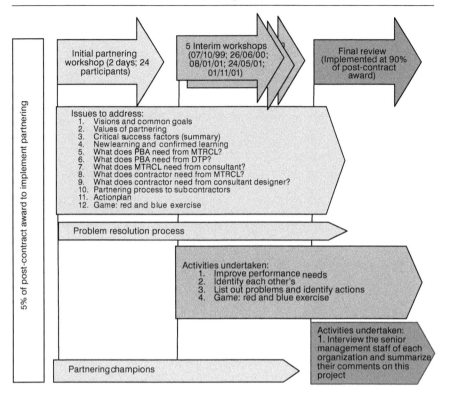

Figure 11.9 The partnering approach and process of MTRCL Tseung Kwan O Railway Extension (TKE) Contract 601 (Hang Hau Station and Tunnels) (adapted from Latham's (1994) Report; source: Chan et al., 2004b)

problems is the day-to-day task, partnering can provide a common platform for beneficial discussions to take place, with a common goal for improving project performance. It is recommended that the adoption of partnering across a wider spectrum of the construction industry greatly assists the achievement of construction excellence.

Case study 16

MTRCL Contract 654 (Platform Screen Doors), Hong Kong – A case study of an underground railway extension project (Source from Chan et al., 2004b) (permission has been obtained for both print and online use from Construction Industry Institute, Hong Kong).

Table 11.7 Key performance indicators (KPIs) for the MTRCL Tseung Kwan O Railway Extension (TKE) Contract 601 (Hang Hau Station and Tunnels) (reproduced from Chan et al, 2004b) (with permission for both print and online use from Construction Industry Institute – Hong Kong)

KPIs	MTRCL TKE Contract 601
1. Construction time	1,176 calendar days
3. Time variation	-5.62%
4. Construction cost	Within budget
5. Injury (accident) rate	30.54/1000 employees
6. No. of complaints received caused by environmental issues	26

Background

The Tseung Kwan O (TKE) project comprises 13 civil contracts, 4 building services contracts, and 17 electrical and mechanical (E&M) contracts. The civil contracts are mainly engineer's design, split geographically among stations, tunnels and a depot. The building services contracts are all design and construct, again geographically split (i.e. stations and ancillary buildings, and a depot). The system-wide E&M contracts are all design and construct, split by discipline, each one covering the whole extent of TKE (MTRC, 2003a).

Work on the TKE was managed for MTRCL by their project division. Construction commenced in late 1998, with opening to the public in the second half of 2002. Since the senior management was convinced from the airport railway experience that adversarial working environments were materially detrimental to the efficient delivery of multidiscipline railway projects, the concept of partnering was initiated. This was followed by the set-up of a senior management steering group to conduct research on partnering in the UK, Australia and Hong Kong (MTRC, 2003b). The steering group's mission was to assess the benefits that could be reaped from partnering and to identify how partnering might be introduced to the MTRCL project division's projects. The conclusions of the group were that the introduction of partnering would improve cost effectiveness, give greater time certainty and result in better communication, more cooperation and quicker problem solving. In 1999, MTRCL decided to adopt partnering for its TKE project. The TKE contractors were invited to participate in a partnering initiative on a voluntary basis although the contract had been awarded on a traditional basis. This was initially supported with varying degrees of enthusiasm by 10 civil contractors, notably with strong support from some leading contractors, and an external partnering facilitator organization (MTRC, 2003b).

Figure 11.10 Outlook of MTRCL Contract 654 – Platform Screen Doors

MTRCL TKE Contract 654 (platform screen doors)

MTRCL TKE Contract 654 is one of the 17 E&M contracts and was composed of the supply and installation of platform screen doors along the whole MTRCL's TKE with five stations. The original contract sum at tender award was approximately HK$131 million (approximately US$16.8 million), with original contract duration of 1,393 calendar days. The key participants included the client and the main contractor. The project was procured by a lump-sum fixed-price design-and-build contract together with an incentive agreement (IA). The mechanism of the IA is developed whereby from an agreed start date, all outstanding works are calculated with risk cost and a gain-share/pain-share arrangement is agreed on with the main contractor. The client and the main contractor share any savings (gains) if the final account is less than the target. Should the final account exceed the target, they share the excess (pain) (MTRC, 2003a).

Partnering approach and process

Figure 11.11 shows the partnering approach and process of MTRCL TKE Contract 654 (Platform Screen Doors) in which there were a total of five partnering workshops. They included one inaugural workshop, one initial partnering workshop, three interim partnering workshops and one final partnering review. The inaugural workshop was mainly to introduce the concept of partnering to the senior management staff of each participating organization. The one-day initial workshop was held at 17% of the post-contract award period with 14 participants. It is of interest to note that unlike the US where the first or inaugural partnering workshop is usually held after contract award but generally before any contract work is initiated, the initial partnering workshop in this project was held after the contract work started. A major reason behind this approach is that partnering was still at an early stage of development in Hong Kong and its implementation was not so widespread when compared with the US and the UK. Moreover, the traditional working relationship between client and contractor is not long-term and is largely on a project-by-project basis. Therefore, many clients may prefer introducing partnering at a later time after they have developed a higher level of mutual trust by working closely together with other parties at the beginning of a project. Four activities were undertaken, including: (1) discussions of visions and common goals; (2) identification of waste and improvement areas; (3) an action plan; and (4) a participation game (red and blue exercise), which included a problem-resolution process and nomination of partnering champions.

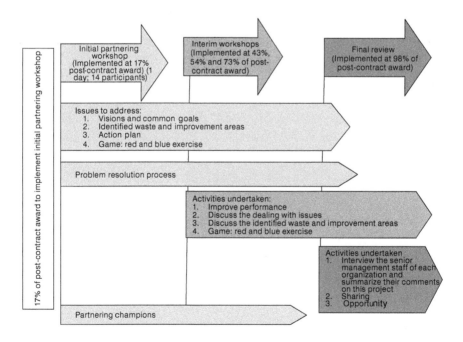

Figure 11.11 The Partnering Approach and Process of Mass Transit Railway Corporation Limited (MTRCL) Tseung Kwan O Railway Extension (TKE) Contract 654 (Platform Screen Doors) [adapted from Latham's (1994) Report] (Source: Chan et al., 2004b)

Results on key performance indicators

The objective KPIs used in this study include: (1) construction time; (2) time variation; (3) construction cost; (4) injury (accident) rate; and (5) number of environmental complaints received. Table 11.8 shows the KPIs for the MTRCL TKE Contract 654 (Platform Screen Doors). It is noted that the time variation was ahead of schedule by 4.95% and the construction cost was within budget. The injury (accident) rate of this project was 57.60/1000, which was lower than the industry average of 85.2/1000 based on the statistics released by the Hong Kong Labor Department in 2002. These KPIs provided evidence that this was a successful partnering project.

Lessons learned

This case study has provided valuable insights into how the partnering culture can be successfully developed through the implementation of IA. Both quantitative and qualitative findings derived from different sources converged to demonstrate outstanding partnering performance achieved in the MTRCL Tseung Kwan O Extension (TKE) Contract 654 (Platform Screen Doors). The underlying basis for this outstanding performance was due to the client's initiative to introduce a novel IA. IA, being similar to target cost contracting (TCC) in principle, facilitated a mutually agreed upon gain-share/pain-share arrangement between the client and the main contractor. Consequently, the three core partnering elements, mutual trust, common goals and commitment were easily achieved under such a mechanism. The implementation of IA underpinned the partnering success on the MTRCL TKE Contract 654 (Platform Screen Doors). Therefore, it is recommended that partnering together with TC contracts such as IA be adopted across a wider spectrum of the construction industry to reap sustainable benefits and achieve construction excellence.

Table 11.8 Key performance indicators (KPIs) for Mass Transit Railway Corporation Limited (MTRCL) Tseung Kwan O Railway Extension (TKE) Contract 654 (Platform Screen Doors) (reproduced from Chan et al 2004b) (with permission for both print and online use from Construction Industry Institute – Hong Kong)

KPIs	MTRCL TKE Contract 654
1. Actual construction duration	1,324 calendar days
2. Time variation	-4.95% (i.e. completion at 69 calendar days ahead of schedule)
3. Construction cost	Within budget
4. Injury (accident) rate	57.60/1000 employees
5. No. of complaints received caused by environmental issues	0

Comparisons of the case studies in the US, the UK, Australia and Hong Kong

Similarities

Most of the case studies in partnering in the US, the UK, Australia and Hong Kong indicated that partnering approach, if implemented appropriately, could assist the industrial practitioners to alleviate the damages brought about by potential obstacles and difficulties and reap the fullest benefits from it. The case studies proved that partnering could induce improved and enhanced relationships amongst project stakeholders by developing common strategies towards win-win outcomes for all parties. By doing so, the project performance could be greatly enhanced. It is important to suitably use different partnering tools, such as holding a series of partnering workshops and regular partnering meetings, development of a partnering charter, selection of appropriate partnering champions, and appointment of experienced external facilitators and training of in-house facilitators. These could assist in developing a strong and continuous partnering spirit, ultimately helping to improve the working relationship amongst the project team members. Therefore, it is highly recommended that partnering be adopted across a wider spectrum of the construction industry to reap sustainable benefits for construction excellence in the twenty-first century. However, it should be noted that partnering is not a panacea and it only provides an avenue where practitioners can communicate better, thus eliminating unnecessary misunderstandings.

Differences and highlights

Based on the above-mentioned case studies in partnering, it can be concluded that partnering was implemented successfully in the USA, the UK, Australia and Hong Kong. However, the reasons for their success are not the same and they should be highlighted.

Unstructured approach

Two case studies in partnering from Hong Kong showed that it is not necessary to adopt a structured approach for successful partnering. Instead, a strong partnering spirit underpins the success of a partnering project no matter whether it adopts structured approach or unstructured one. Structured approach with strong partnering spirit, however, makes a partnering project more successful because it can provide extensive benefits to the industry across different sectors.

Extended partnering

A case study in partnering from Australia indicated that an extended partnering approach could create stronger and more efficient and effective teamwork and cooperation between contracting parties, thus resulting in better project performance in terms of time, cost, quality, safety and environmental performances.

Partnering with value engineering management

A case study in partnering from the US revealed that partnering together with value engineering management could assist in enhancing the working efficiency amongst the parties, thus achieving better project performance.

Partnering with guaranteed maximum price

Two case studies in partnering from Hong Kong were procured by a GMP contract. It has been observed that GMP may be a good mechanism contributing to partnering success because under this mechanism, a client and a main contractor are needed to work as a team to determine many vital events, such as construction method, programme, pricing, detailed breakdown of works, and consenting to preliminaries and conditions of contract. This entails the main contractor to release all the back-up data to team members. Under this working environment, it is easier to develop trust amongst the parties.

Partnering with incentive agreement (a kind of target cost contract)

Two case studies in partnering from Hong Kong adopted partnering approach with IA and such a combined approach is proved to be able to assist in achieving outstanding partnering performance. A major rationale behind is that IA, being similar to target cost contract in principle, facilitates a mutually agreed gain-share/pain-share arrangement between a client and a main contractor. Under this situation, mutual trust, common goals, and commitment, three core elements of partnering, can be more easily achieved.

Chapter 12

Case studies in alliancing

This chapter shows how alliancing is implemented in two different nations. A total of 12 case studies in alliancing adopted in the UK and Australia are selected for demonstration purposes.

Case study 1

Development of a pan-European co-marketing alliance – A case study of British Petroleum (BP)-Mobil (Extracted from Robson and Dunk, 1999) (permission has been obtained for both print and online use from RightsLink Copyright Clearance Centre).

Background

Since the mid-1980s, strategic alliance has received much attention from academic researchers in the fields of international marketing and strategy. There are two vital issues in this field. First, many managers have encountered enormous difficulty to set up and manage inter-organizational entities. It is found that many empirical searches on international alliance report unsatisfactory performance (Beamish and Delios, 1997). Second, there is a paucity of research regarding the establishment and development of effective international strategic alliances in marketing specifically. Achrol et al. (1990) stated that the managerial difficulties posed by co-marketing alliances have been neglected by academic researchers in the collaborative strategy field.

The aim of this case study is to provide business practitioners who are interested in formulating and implementing a successful co-marketing alliance strategy with valuable insights by exploring the developmental process of a successful strategic alliance through a case study of BP-Mobil, with interviews with senior executives from the alliance entity. By doing so, business practitioners can get three major benefits from the case study. First, it helps identify how to deal with the implicit organizational complexities that are linked negatively to the effective establishment and management of large cross-border alliances. Second, it provides insights into the difficulties in relation to the process of developing concurrently an alliance strategy in marketing in many European markets. The third benefit is that the study reinforces the validity of the statement 'the end justifies the means' with regard to achieving a solid competitive position in a stagnant business environment via a radical deviation from the strategic status quo.

Framework of analysis

There are two perspectives on developing successful co-marketing alliances used to evaluate the case study (BP-Mobil): symbiotic marketing and inter-organizational exchange theory. Symbiotic marketing edifies the synergistic benefits that firms may gain by collaboratively exploiting growth opportunities while maintaining their distinct identities (Adler, 1966; Varadarajan and Rajaratnam, 1986). Under this framework, Varadarajan and Rajaratnam (1986) developed a system for planning co-marketing agreements in which the combination of partners' resources and programmes increases the market potential of each firm (Adler, 1966). A successfully pursued symbiotic marketing strategy will supplement capabilities, achieve economies of scale and scope, capitalize on synergistic relationships among products, overcome constraints imposed by resource limitations, lower business risks and better satisfy customer requirements (Varadarajan and Rajaratnam, 1986). By adopting this approach, BP-Mobil was assessed based on strategic growth potential and the presence of favourable organizational factors. The second theoretical framework, inter-organizational exchange theory, viewed that firms are driven to collaboration by a need to exchange resources because they have narrow functional capabilities and operate in an uncertain environment in which resources are scarce (Pfeffer and Salancik, 1978).

Overview of BP-Mobil's successful evolution

On 29 February 1996, BP and Mobil announced that they intended to tie-up downstream operations in the refining and marketing of fuels and lubricants across Europe, subject to European Union (EU) approval. The EU Competition Commissioner authorized this planned link in August of that year, and measures were immediately taken by the new allies to implement their radical conception: the creation, through collaboration, of a first-tier player out of two oil firms whose individual prospects in the oversupplied European market are limited. After alliance, BP-Mobil immediately became the second largest oil marketing company in Europe.

BP and Mobil were amongst the first to rationalize their downstream operations in response to the chronic over-capacity within the sector. Before the commencement of BP-Mobil, both parent firms made major refining cutbacks. In fact, the alliance strategy is the continuation of a series of substantive cost-reduction measures on the part of both parents. This landmark partnership reduces further duplication, improves refinery location and drives down costs. BP are committed to creating a competitive advantage through learning and developing, and consider the strategic alliance approach to be a major part of creating distinctiveness. The executives of BP viewed that strategic alliance facilitates breakthrough thinking and value maximization and should be used for competing within a firm's core markets and technologies (Prokesch, 1997).

Table 12.1 summarizes the symbiotic nature of the BP-Mobil co-marketing alliance. The dimensions adopted were proposed by Varadarajan and Rajaratnam (1986) to permit the identification of individual co-marketing alliances from a wide variety of alternative inter-firm marketing relationships in existence.

Evaluation of BP-Mobil

Strategic perspective

Since both BP and Mobil continuously strive to give greater value to shareholders, it is vital that BP-Mobil cancels out the cutbacks by capturing an enlarged market share. In this context, a major benefit of this alliance would

Table 12.1 Scope of BP-Mobil's co-marketing alliance (reproduced from Varadarajan and Rajaratnam, 1986) (with permission for both print and online use from RightsLink Copyright Clearance Centre)

Dimension	Description
Time frame	It is the intention of both strategic partners to develop a continuing business relationship consistent with their long-term growth objectives for Europe.
Proximity	Close-working: the strategic partners operate as if a single pan-European business entity. The European fuels and lubricants personnel of each side have been fully integrated, and the alliance's fuels and lubricants management teams share office facilities.
Number	Simultaneous multiple: the BP-Mobil co-marketing alliance is consisted of separate, complementary, international alliances for the fuels and lubricants business.
Level	Organizational: both international alliances span two oil business value chain activities, namely refining and marketing.
Focus	Product offerings of both partners: the goods of both firms are central to the symbiotic relationship.
Scope at the marketing function level	Broad: this integrated marketing strategy is an opportunity for leveraging in all four components of the marketing mix.

be the increased potential for product development-led growth by means of Mobil's great R&D strength, the improved human and financial resources, and superior marketing intelligence from an expanded distribution network. This positive outcome is good for the future and would require a prolonged period of work and mutual cooperation. BP and Mobil anticipate that once fully operational, the critical mass partnership will have the asset base to command currently served markets and forge inroads into new geographic markets. BP-Mobil's marketing strategy will bring to bear a broad product assortment on national custom, in order that local needs and desires are better served.

BP-Mobil adopted a complicated rebranding strategy that is aimed at minimizing consumer confusion and maximizing marketplace success, while protecting the brand equity of both parent firms. In lubricants, BP-Mobil has retained all three brand names (i.e. Mobil, BP, Duckham's) so as to capitalize on each strong identity. This has led to the appointment of separate advertising agencies to deal with the Mobil, BP and Duckham's accounts. In contrast, a uniform fuel offer has been made under the re-profiled BP brand name, involving the transformation of Mobil's 3,300 forecourts into BP's distinctive green and gold livery. However, the alliance logo, which depicts the idea that two separate identities have come together to provide additional value, is displayed clearly on the signage.

Factors fostering symbiosis

From the symbiotic marketing standpoint on emerging growth opportunities, robust cooperation between two or more firms is the most desirable strategic option where there are significant resource complementarities; combined strengths lie in a number of key success areas; and operations are exposed to considerable product-market risks. Resource conditions are conducive to this alliance strategy because the division of resources is straightforward and their utilization is efficient. Cost savings from the elimination of duplicated services, synergies in marketing and distribution, and economies of scale via pooled purchasing volumes are expected to reach \$400-\$500 million per year by the end of 1998 and continue to accumulate into the next century. This high level of operational compatibility is present equally in the lubricants and fuels businesses. The complementary and compensatory strengths and weaknesses of BP and Mobil are suggestive of the co-marketing alliance's long-term viability. By combining downstream operations, BP and Mobil have improved their standing vis-à-vis product-market risk in three ways. First, individual investment outlays have been reduced for the risk-fraught expansion eastwards. Second, both parent firms were forced to look inwards and undertake additional steps to rationalize and upgrade their refining operations. Third, there was an extended pan-European service stations network for BP-Mobil.

Project management

When a strategic alliance strategy is formulated, experienced firms try their best to observe and manipulate the determinants of inter-partner conflict, so as to build stability into the continuous business relationship (Parkhe, 1991; Lyles and Salk, 1996; Ding, 1997). Moreover, these firms avoid engaging in alliance projects in which there is definite potential for dysfunctional conflict (Gomes-Casseres, 1999). The extent to which BP-Mobil has embraced this prescript and is vulnerable to the destabilizing effect of the main conflict drivers: (1) power imbalance; (2) managerial imbalance; and (3) inter-partner diversity were assessed. The assessment concluded that BP-Mobil was well managed.

Lessons learned

The BP-Mobil experience provides business practitioners with valuable insights into the establishment and management of such a symbiotic business entity. Inter-organizational exchange theory also fully supports this alliance strategy, insofar as the formative process behind BP-Mobil limits conflict pertaining to the individual interests that the integration embodies, and financially BP-Mobil constitutes a sound investment. In fact, the success of this case study draws some lessons learned in terms of: partner selection; venture design; and venture management.

Partner selection

Prospective parent firms should try hard to secure a high degree of inter-partner synergy, because without this there will be little basis for the cultivation of a harmonious business relationship. In this regard, a first prerequisite would be to ensure that the growth objectives of partners and long-range visions for the sector(s) in which the new business would operate are fully compatible. Second, it is vital that prospective parents utilize partner option to prevent future conflict pertaining to the management systems absorbed into the new management structure.

Venture design

During the chaotic transition period, there are three design factors that play a pivotal role in bypassing impediments and accelerating process. First, it is important that partner firms jointly set up project-based teams to overcome situation-specific barriers to the co-marketing alliance's effective functioning, along with a committee with the seniority and diplomacy to tackle disputes and maintain partner rapport. Second, partners should understand the operational value of custom of best practice and structure the alliance opportunity with humility and pragmatism. The third design factor to be used concerns defining the confines of the alliance entity properly. The damaging effect of any ambiguity concerning the alliance's position and role vis-à-vis parent firms within their inter-organizational networks must not be overlooked.

Venture management

Two managerial factors essential to the effective management of BP-Mobil may be employed successfully in other co-marketing alliances. First, it is vital that the leadership of each parent firm gives discernible signs of long-term commitment to their specific alliance strategies and co-collaborators, as well as to the alliance approach itself and all their representatives in strategic alliances. Second, it is crucial that venturing firms achieve and sustain full inter-partner cooperation by means of the development of a relational contract.

Case study 2

Development of effective alliance partnerships – Lessons from a case study of OilUK (Extracted from Hipkin and Naudé, 2006) (permission has been obtained for both print and online use from RightsLink Copyright Clearance Centre).

Background

Alliances are often used to enhance organizational learning, technological leadership and knowledge-based capability. Close interaction between partners can complement internal development and permit faster access to new technologies located beyond the boundaries and abilities of an individual firm. However, many alliances are quite static and their evolution is predictable. This is a longitudinal study of an alliance case whose initial strategic objectives were overtaken by innovative developments that finally led to the creation of an unusual knowledge-based configuration. The study investigates this evolution using three themes, including: (1) how strategic activities evolve alongside technology; (2) how expertise influences partnership development; and (3) how governance of an alliance influences outcomes. The case study shows how learning and innovation determined the strategic direction of the case alliance, creating occasions for more opportunistic experimentation. It is also revealed how partners gained influence through their capabilities and knowledge contributions, and how the alliance was affected more by partner expertise than by initial strategic intentions.

Approach to the case study

This case study examines an extraordinarily interesting alliance development. The case study uses qualitative methods, causal analysis, observation and interviews to document what happens and the reasons behind them. As the data for this study were drawn from interviews, it has to be satisfied that the details and the inferences are indeed genuine observations and a true reflection of a situation, and that through analysis of the interview data, it is possible to learn about the social world beyond the interview context. A shortcoming of past academic studies of alliances is that they were frequently examined at a single point in time (usually in the formative stages) without following the alliance through its life or even a business cycle as strategic and environmental conditions change. However, this alliance case was studied over three years.

The major alliance partner, OilUK, is the lubrication division of a UK multinational company, a supplier in a sector noted for rapid competitor imitation of new initiatives. Firms in this sector have long realized that a low-cost strategy for the sale of lubricants is inadequately profitable. Value to the product has been added by offering surveys of lubrication requirements for all plant and equipment in an installation. This tactic has proved to be relatively effective, as the recommended lubricant is invariably only sold by the supplier undertaking the survey.

Figure 12.1 shows that there are four developmental stages of the alliance. At stage 1, OilUK offered a fully out-sourced service to clients in the UK, or through a subsidiary or distributor elsewhere. The strategy of OilUK was global expansion through a range of innovative lubrication services. Before launching a global service, a pilot project was to be undertaken by OilUK's South African subsidiary (OilSA) to establish whether the planned services should be developed internally or through an alliance. In spite of early difficulties with the development of a simultaneously complicated and new service, two South African clients took part in the venture as alliance partners. At the beginning, OilUK recognized that there may not be a critical mass ready to absorb the technology in South Africa. Nevertheless, the exercise would be a valuable learning experience for expansion into other markets. Clients were to be offered an oil analysis condition monitoring service that would assist them to predict failures, and reduce the risk to OilUK of having to pay penalties for lubrication-related breakdowns.

The next development was to enlist an expert condition monitoring partner (referred to as ConMon), to analyse and interpret oil analysis and machine performance data. A loose alliance was formed between OilUK, OilSA, and ConMon early in 1998. Two OilSA project managers coordinated day-to-day activities of the pilot studies. Two opencast mine clients (referred to as Morafo and Umgodi, Sotho and Zulu words for 'mine') agreed to make

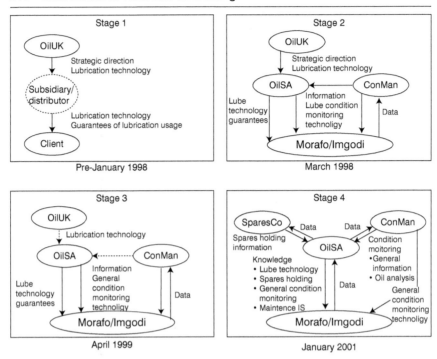

Figure 12.1 Stages of alliance development (solid lines show communication often followed between partners; dotted lines imply a looser exchange of ideas or communication (formal channels may exist, but were not always followed) (reproduced from Hipkin and Naudé, 2006) (with permission for both print and online use from RightsLink Copyright Clearance Centre)

use of the condition monitoring service, and effectively became partners in the alliance.

By the end of 1998, engineers at Morafo recognized the potential for expanding conditioning monitoring facilities, and installed equipment to measure bearing vibration and compressor performance. These data were sent to ConMon for analysis, and some potentially expensive failures were prevented. Within a few months relationships were changing in that most communication took place directly between the mines and ConMon, with OilSA being largely sidelined. At stage 3, ConMon provided a wide variety of analyses, and the mines took suitable action, without referral to OilSA. By this stage, OilUK was effectively barred from alliance activities.

OilSA introduced a further service in August of 2000, in conjunction with a second UK company, SparesCo, which offered advice on maintenance spares holding policies. With previous history now held by OilSA, and deterioration rates provided through ConMon, SparesCo was able to recommend stockholding policies to reduce investment in maintenance spares. Stage 4 shows the situation where OilSA had largely taken control of most data and information transfers between alliance partners.

Lessons learned

There are a number of lessons learned from this case study and they can be classified under three themes, including: (1) evolution of strategic objectives; (2) contribution of expertise to strategic partnership; and (3) development of alliance governance.

Evolution of strategic objectives

First, partners explore the unknown and rely on imaginative capability-based expertise for strategic direction. Second, objectives evolve through learning, experimentation and innovation. Third, strategies in high-technology partnerships can evolve in a significant and highly positive way through learning, experimentation and innovation. Firms take advantage of unanticipated opportunities and tolerate failures. Managers recognize obsolescence and technology shifts.

Contribution of expertise to strategic partnership

Knowledge-based alliances depend on the unique capabilities of individual partners. Contribution of partners fluctuates as one partner's expertise is superseded by that of another. Partner with static strategy risks being left behind.

Development of alliance governance

Intuitive beliefs initiate alliance, but loose governance enables leverage of resources and capabilities. Governance should foster a learning culture and danger exists that a strong strategic centre becomes core rigidity. Strategy formulation, management of partners, and technological/knowledge development progress are in parallel.

Case study 3

Boyne Island Development project, Australia – A case study of a large-scale private sector project.

Background

This was a private sector project adopting alliance contract in Australia. The project began in 1992 and was completed in 1993. The contract sum was approximately AU$1.5 billion and there were about 1,000 workers involved in this project. The client was Comalco Ltd and the main contractor was Bechtel Ltd (an American company).

Alliancing approach and process

At first, the client and the main contractor mutually designed the contract and defined the terms of condition. Afterwards, they passed it for contractors to follow and then they set some KPIs to meet, such as time, cost, quality, safety and innovation.

Results of the interviews

Major reasons for adopting an alliance contract

The major reasons for adopting an alliance contract in this project were to avoid adversarial working relationships amongst different parties and to share risks with them because the design was incomplete and there were lots of uncertainties and risks.

Perceived major benefits and critical success factors for adopting an alliance contract

The interviewee perceived two major benefits when alliance contract was adopted in this project. First, there was a risk cap in this project, thus making profits known to each party and second, the working relationship with other parties was much better. The critical success factors for adopting alliance contract in this project were to deliver the project on time and to meet the client's satisfaction.

Relationships in the alliance contract

The working relationships between parties was very good in that industrial practitioners were willing to work as an integrated and dedicated team.

Major difficulties in alliancing implementation

The interviewee viewed that it was difficult to keep the KPIs right because the contractors focused it incorrectly, but this problem was resolved smoothly by the adjustment of contractor representatives.

Effect of construction culture on alliancing practice and performance

The interviewee stated that the labour cost was very high so clients would like to keep employees to a minimum. The construction approach was dynamic and changing quite fast because technology changed very quickly. People were more concerned about safety. Under this culture, it was more flexible for contractors to put their inputs into the design because the labour cost was quite high. The culture promoted alliance contracts and helped this project much because there were lot of uncertainties and risks.

Effect of organizational culture on alliancing practice and performance

The interviewee stated that there were fewer procedures and rules for his colleagues to follow so that it was more flexible for them to work in the company. This culture supported alliance practice in this project because people could work more flexible, thus inducing more innovative ideas and problems were solved quickly. It was also very favourable to the alliancing performance because it assisted in building up an integrated team and senior management staff were willing to commit and they were more adaptive to new ideas.

Lessons learned

The lessons learned from this successful alliancing project could help the practitioners to mitigate the harms brought about by possible difficulties, and maximize the benefits gained from implementing alliancing concept. It is observed that an alliancing approach is especially suitable for those construction projects that are complex in nature, with high risks and great uncertainties inherent in the project. By using this approach, this helps to fundamentally change the motivation and dynamics of the relationship between alliance members so that it could lead to better project outcomes through mutually developed strategies towards win-win outcomes for all stakeholders. It is concluded that project alliancing, when implemented properly, can generate a good practical model for different parties to work more effectively and efficiently, thus eliminating potential project risks and enhancing certainties in implementing the project.

Case study 4

Sewerfix Alliance project, Australia – A case study of a medium-sized infrastructure project.

Background

This was a wastewater infrastructure project which upgraded sewer water plumb in Sydney. The project began in late 2001 and was completed in mid 2004. The contract sum was approximately AU$230 million and there were about 300 to 400 participants involved in this project. The client was Tenix Alliance Company, the main contractor was Bovis Lend Lease, and two main consultants were Ch2M Hill and Sinclair Knight Merz Ltd.

Alliancing approach and process

At first, a call for proposals was made through public advertisement. Four parties were then short-listed and they attended an interview meeting during a one-day workshop. After assessment, two parties were selected to attend a two-day workshop separately. Finally, the most capable party won the project. The whole process took about ten weeks.

Results of the interviews

Major reasons for adopting an alliance contract

The major reasons for adopting an alliance contract in this project were that there were many uncertainties and much risk in the project. In addition, the client would like to use fast-track approach to procure this project under tight schedule.

Perceived major benefits and critical success factors for adopting an alliance contract

The interviewee perceived that there were several major benefits (Figure 12.2), which included: (1) outstanding success for the client in terms of time, cost and quality; (2) profitable for all parties; (3) good working relationships; (4) good safety performance; and (5) excellent relationships with the community.

There were two critical success factors for implementing an alliance contract in this project. These included: capable project team was selected not based on the lowest price but other non-price selection criteria; and high level of client's involvement.

Relationships in alliancing

The interviewee reckoned that the project team was highly integrated at all level.

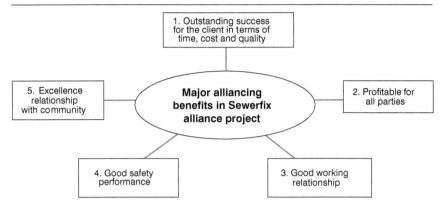

Figure 12.2 Major benefits of alliancing for Sewerfix alliance project

Major difficulties in alliancing implementation

Although there were a lot of difficulties encountered in this project, they were resolved smoothly because different parties were willing to work together as an integrated team.

Effect of construction culture on alliancing practice and performance

The interviewee viewed that the construction culture in Australia has several characteristics, encompassing: (1) people work more equally; (2) workers are strongly protected by the unions; and (3) highly adversarial working relationships in the building industry but better working relationship in infrastructure, civil, construction and mining.

Under this culture, the vertical integration was good (from top to bottom) and the working relationship between parties was not adversarial.

Effect of organizational culture on alliancing practice and performance

The interviewee opined that the organizational culture in his company was characteristic of: (1) highly performance-oriented; (2) very 'flat' structure: fewer layers; and (3) very friendly and supportive. However, this culture did not affect alliance approach in this project much.

Lessons learned

The lesson learned from this successful alliancing project is that selection of capable project teams is of utmost importance to alliance success. It is recommended that project team members should not be selected mainly based on the lowest tender price. Instead, other non-price selection criteria should be evaluated with greater importance.

Case study 5

Windsor Road Alliance project, Australia – A case study of a small-sized infrastructure project.

Background

This was a road infrastructure project and it began in February 2005 and was completed in July 2006. The contract sum was approximately AU$100 million (the contract sum of the whole project was around AU$300 million and the full name was Windsor Road Upgrade Project) and there were about 120 participants involved in this project, of which 20 came from the management team and 100 from the working level. There were four major parties involved in the project, encompassing the client (Road and Traffic Authority), the head contractor (Leighton) and two main consultants (Maunsell and Coffey) and they formed a consortium.

Alliancing approach and process

At first, the client used an alliance contract in this project. Then, he recruited an independent alliance facilitator to assist him to employ the most capable project team through drafting tender documents, and conducting interviews and alliance workshops. Four project teams submitted their proposals and they were assessed through interviews and workshops. Two capable teams were selected to enter the final round (Leighton and Abi Group) of interview. Finally, Leighton won the project. After that, all the parties drafted the alliance contracts together (they agreed with the schedule, budget and other project details).

Results of the interviews

Major reasons for adopting an alliance contract

The interviewee stated that the reasons behind for adopting an alliance contract in this project were: the project was very complex (highly built-up suburban residential area); and the schedule was very tight.

Perceived major benefits and critical success factors for adopting an alliance contract

The interviewee viewed that the project team enjoyed a number of benefits when implementing alliance contract (Figure 12.3), including (1) building up a long-term business relationship; (2) building up good reputation; (3) good profits; (4) satisfaction of all parties; and (5) future business promotion.

The CSF for adopting alliance contract in this project was that alliancing parties truly abided by the alliancing spirit (everyone behaved in the right manner).

Relationships in an alliance contract

The working relationship between the client and the project team was excellent because the traditional client-contractor confrontational relationship did not exist (no claims and disputes).

Major difficulties in alliancing implementation

The major difficulties encountered when alliance contract was adopted included: (1) the traffic was very heavy; (2) complex underground utilities; and (3) difficult to plan construction activities. Nevertheless, the problems were resolved by consultation with the community; good traffic directions; and good liaison with utilities companies.

Effect of construction culture on alliancing practice and performance

The interviewee perceived that the construction culture in Australia was traditionally 'can do', meaning that nothing is impossible. People faced challenges in a positive way and difficulties could be overcome effectively. This culture was very helpful to alliance practice in this project because it

Figure 12.3 Major alliancing benefits for Windor Road alliance project

encouraged people to work closely at each stage. When problems arose, all the parties focused on finding quick solutions so that the project could be completed on time.

Effect of organizational culture on alliancing practice and performance

The interviewee opined that the organizational culture in his company was also 'can do'. He added that his company was famous for delivering projects with high level of efficiency and on time. Since there were a lot of capable people in his company, their expertise and practical knowledge could contribute to the project outcomes in terms of time, cost and quality.

Lessons learned

This case study confirmed that alliancing approach is particularly appropriate for projects facing tight schedule, complex situations and high risks with high levels of uncertainty.

Case study 6

Northside Storage Tunnel project, Australia – A case study of a large-scale infrastructure project.

Background

This was a large-scale infrastructure project and it began in 1997 and was completed in 2000. The contract sum was approximately AU$450 million and there were about 300 workers involved in this project. The key participants included the client (Sydney Water Corporation); the main contractor (Transfield Contractor); and a main consultant (Connell Wagner).

Alliancing approach and process

The alliancing approach in this project was implemented by non-cost selection approach. Under this method, some suitable contractors were invited and the most capable contractor (Transfield Contractor) was selected based on its own capability, professional attitude and resource management, but not based on price (so-called non-financial ability). After selecting the Transfield Contractor, the owner needed to calculate the target out-turn cost (TOC), which was calculated based on the 3-limb approach. The limb 1 included direct cost and site overhead cost. Limb 2 encompassed head office overhead cost and marginal cost. And limb 3 included cost involved in gain-share/pain-share arrangement.

Interview results

Major reasons for adopting an alliance contract

The major reasons to adopt an alliance contract in this project were to meet a very tight schedule and to deal with an unclear scope of work. (The interviewee opined that it was more suitable for a project team to adopt an alliance contract for complex projects because there were many uncertainties for managing them. In contrast, it was better for a project team to use partnering approach for less complex projects because time and cost were more predictable.)

Perceived major benefits and critical success factors for adopting an alliance contract

The perceived major benefits for adopting an alliance contract, as stated by the interviewee, included: (1) solving adversarial conflict problems; (2) resources were fully utilized (no waste); (3) good working relationship between parties (a good team); and (4) increasing the reputation of all companies.

The CSFs for adopting an alliance contract were (1) finished by a fixed time; (2) environmental sensitivity; and (3) top management commitment.

Relationships in an alliance contract

The working relationship between parties was very good and some practitioners finally merged in one team.

Major difficulties in alliancing implementation

Since ground condition was bad and the cost involved in solving this problem was beyond the planned budget, many companies criticized that this method was not value for money. To solve this problem, the project

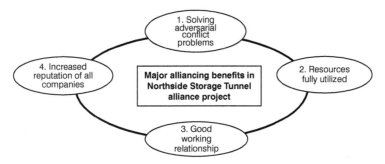

Figure 12.4 Major alliancing benefits for Northside Storage Tunnel alliance project

manager tried to analyse it rationally. However, some companies did not satisfy his explanations (but this was not a very serious problem).

Effect of construction culture on alliancing practice and performance

The interviewee believed that traditionally, the working relationship between different stakeholders was very aggressive, often resulting in adversarial relationship with unproductive working environment. But now practitioners are much more interested in building up relationships through cooperation and communication. They are willing to change the traditions and accept new ideas to procure their projects.

This culture supported alliancing approach in this project because it encouraged practitioners to be willing to change the traditions and accept innovative management methods. In fact, different parties were willing to cooperate with open and honest manner.

The performance of this project was very good under this culture (team members were willing to work cooperatively). It is clear that to be successful, team members had to be honest, open and constructive.

Effect of organizational culture on alliancing practice and performance

The interviewee stated that the organization culture in his company was very dynamic (willing to change to cope with new environment). This culture was very supportive to alliancing practice in this project because stakeholders were more willing to accept changes. It was also very supportive because senior management staff was more willing to commit and they were more adaptive to new ideas.

Lessons learned

This case study showed that non-cost selection approach of project team members is crucial for successful alliance. And it is suggested that contractor ought to be selected based on its own capability, professional attitude, and resource management, but not based on price.

Case study 7

Cultural Centre Bus Way project, Australia – A case study of a small-scale infrastructure project.

Introduction

This was a small-scale road infrastructure project and it began in 2003 and was completed in 2004. The contract sum was approximately AU$10 million and there were about 80 people involved in this project, of which 5 were from the client, 20 from consultants and 55 from contractors. The key participants included the client (Queensland Transport Government) and the main contractor (Abigroup Contractors).

Alliancing approach and process

At first, the contractor actively discussed with the client and explained that it was not good to use traditional approach to procure this project because although the contract sum of this project was low, it was very complex because it was involved many stakeholders, especially the community. The client was finally persuaded by the contractor and agreed to use alliance contract with a cap amount (risk sharing amongst different parties). Afterwards, it brought all the stakeholders together through alliance workshops and meetings. Then, they formed a good team and decision-making processes were quick and problems could be resolved rapidly.

Interview results

Major reasons for adopting an alliance contract

The interviewee explained that there were both political and economic reasons for adopting an alliance contract in this project. For political reasons, the project had to be completed at a fixed date (the end day would not be changed), while for economic reasons, the budget was inadequate and beyond what they initially expected (therefore there must be a risk cap in this project).

Perceived major benefits and critical success factors for adopting an alliance contract

The perceived major benefits for implementing alliance contract encompassed: (1) good reputation; (2) meeting the needs of end users; (3) satisfaction of politicians, stakeholders, bus users and the client; (4) sharing savings amongst different parties (reasonable profits); and (5) from the government point of view, showing that alliance contract really works.

The CSFs for adopting an alliance contract in this project were: (1) commitment of senior management from all parties; (2) excellent team; and (3) community support.

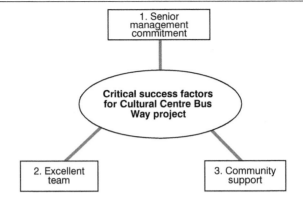

Figure 12.5 Critical success factors for Cultural Centre Bus Way project

Relationships in an alliance contract

The working relationship between parties was very good in that they formed an integrated team and they trusted and respected with each other.

Major difficulties in alliancing implementation

A major difficulty was related to design and construction in that the roof was too heavy and project stakeholders had different opinions to deal with this problem, but being an integrated team, they solved this problem efficiently at last.

Effect of construction culture on alliancing practice and performance

The interviewee considered that the construction culture in Australia was more relationship-based than before – people were more cooperative and fewer arguments arose between parties (less adversarial working relationship amongst stakeholders).

If the project started five years before, the client would not be persuaded to adopt alliance contract because the construction environment was still confrontational. But in the years just preceding the project, the culture changed to be more cooperative so that the client was willing to accept innovative approach to procure the project. It was also very positive in terms of cost and time performance.

Effect of organizational culture on alliancing practice and performance

The interviewee mentioned that it was quite difficult for him to comment on the organizational culture in his company in detail, but he would like to summarize that it emphasized relationships, and colleagues tended to be trustful with each other and have an open attitude. In addition, commitment of senior management was strong. This culture indeed helped internal integration within the company because colleagues at all level believed that alliance contract could make this project a great success. Since top management commitment was strong and there was a dedicated team, this project was delivered with very good performance, especially meeting the tight schedule and budget.

Lessons learned

This case study has provided valuable insights in that alliancing approach can also be suitable for small-scale projects once the project is complex with high risk and great uncertainty.

Case study 8

Woolgoolga Water Reclamation Plan project, Australia – A case study of a small-sized infrastructure project.

Background

This was a small-scale infrastructure project which upgraded a water reclamation plant. The project began in August 2003 and was completed in January 2005. The contract sum was approximately AU$18 million and there were about 43 key participants involved in this project, of which 3 were from the client, 12 from the main contractor, 12 from consultants and 16 from key subcontractors. The client was Coffs Harbour City Council, the main contractor was Abigroup Contractors, and a main consultant was GHD Consultants Ltd.

Alliancing approach and process

At first, the client advertised publicly for an alliance facilitator and SPAN (the interviewee's company) was appointed. Then the alliance facilitator assisted the client to define the scope of work and to identify commercial terms. After doing these works, the alliance facilitator assisted the client to draft all the documents. Then, public advertisement to recruit the most capable team was made. Six teams applied for this project and they were assessed based on non-price selection criteria. Finally, the most capable company and project team won the project.

Interview results

Major reasons for adopting an alliance contract

The major reason for adopting an alliance contract in this project was that the client would like to have more involvement in this project so as to make the best use of the expertise of its professional staff (because they had rich experience in operation and maintenance of their wastewater treatment plants as well as community and stakeholder management).

Perceived major benefits and critical success factors for adopting an alliance contract

The interviewee stated that there were four perceived major benefits for adopting an alliance contract. These included: (1) repeated work; (2) built-up reputation because this was the second alliancing project in the region of Mid Coast of New South Wales; (3) the client was happy to get satisfactory outcome (the project was completed within time and budget (under very tight schedule and budget); and (4) attracted high-quality skilled providers to the region.

The CSFs for adopting an alliance contract in this project encompassed: (1) capable project team (consortium); (2) the alliance facilitator proactively dealt with the people and relationship; and (3) the client team was very supportive to the alliance contract.

Relationships in the alliance contract

The project team was fully integrated and they worked well under alliance principles. When conflicts arose, members were willing to resolve it openly.

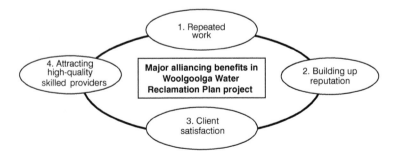

Figure 12.6 Major alliancing benefits for Woolgoolga Water Reclamation Plan project

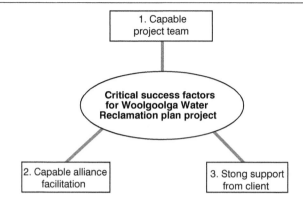

Figure 12.7 Critical success factors for Woolgoolga Water Reclamation Plan project

Major difficulties in alliancing implementation

A major difficulty was shortage of subcontractors in the region. It was resolved by recruiting subcontractors from other areas due to good network between parties. The design team was based in Brisbane, a five-hour drive away. Ensuring that the designers made regular site visits and attended major onsite meetings helped to break down the distance barrier.

Effect of construction culture on alliancing practice and performance

The interviewee perceived that the construction culture in Australia is characteristic of: (1) people like to do a project as quick as possible; (2) they try to fully utilize local resources; and (3) they desire to do good quality work.   Under this culture, alliance contract was easily implemented because parties would like to do the work once only. They were proud of what they did.

Effect of organizational culture on alliancing practice and performance

The interviewee mentioned that the organizational culture in his company is featured of: (1) very collaborative culture; (2) colleagues are fairly open; (3) they are willing to share opportunities; (4) they look for the best solution; and (5) they share the work around. Under this culture, strong alignment between parties was formed and more innovations were created and they were credible to the client and the project team.

Lessons learned

This case study has provided useful insights in that the appointment of a capable external facilitator may be very useful for the alliance success because the capable external facilitator can assist the client to draft well-suited alliance contract and select the most capable project team.

Case study 9

Museum of Tropical Queensland project, Australia – A case study of a small-scale public sector project.

Background

This was a small-scale public sector project with the contract sum of approximately AU$18 million. It began in 1997 and was completed in 1999. There were about 150 participants involved in this project, of which 2 were from the client organization, 48 from consultants, and the remaining from contractors. The client was Queensland Museum Board and the main contractor was Leighton Contractors?

Alliancing approach and process

At the beginning, it was the interviewee who drafted the proposal and persuaded the Queensland Museum Board to adopt alliance contract in this project. After getting approval from the Board, team members were recruited through public advertisement. There were some requirements to be recruited as a team member, and one of them was to be willing to cooperate with other companies with open book manner.

Interview results

Major reasons for adopting an alliance contract

The major reason for adopting an alliance contract in this project was simply to ensure that the project could be completed on time, within cost and of a high quality.

Perceived major benefits and critical success factors for adopting an alliance contract

The perceived major benefits were that the project was able to be completed within time and on budget, and the reputation for the client was also enhanced. The three CSFs for using alliance contract were: (1) excellence in design (because team members could communicate better); (2) good sourcing materials (the cost was lowered due to good sourcing materials); and (3) harmonious working relationship with the end-users (end-users here referring to the director of client and his staff). The working relationship between parties was very good and good social contacts were established.

Major difficulties in alliancing implementation

At the very beginning, subcontractors were not familiar with alliance contract and they hesitated to agree with this kind of formal contract. Finally, all capable subcontractors remained unchanged but all incapable subcontractors were dismissed.

Effect of construction culture on alliancing practice and performance in this project

The interviewee described that the working relationship between different stakeholders was very adversarial before 1990s. The major reasons behind were shortage of professional staff and difficult to handle construction management technique. Because of these reasons, industry practitioners were keen on finding better ways for improvement (willing to change and accept new ideas). After 1990s, the working relationship amongst different parties became less adversarial when alliance contract was introduced.

This culture supported alliancing approach in this project because it encouraged practitioners to be willing to change the traditions and accept innovative management methods. In fact, different parties were willing to cooperate with open and honest manner.

The performance of this project was very good under this culture (team members were willing to work cooperatively) and they would like to benchmark this project to enhance their reputation.

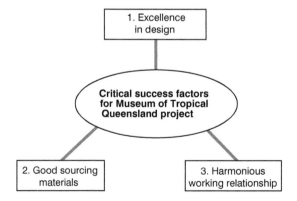

Figure 12.8 Critical success factors for Museum of Tropical Queensland project

Effect of organizational culture on alliancing practice and performance

The interviewee stated that the organization culture in his company was to identify the best consultant and contractors to deliver the project. This culture was very supportive to alliancing practice in this project because the quality assurance system of the interviewee's company was very high (able to certify both external processes and internal processes very well), thus making the project to be value for money. In addition, it was also very supportive to the alliancing performance because it helped enhance the quality of project, thus building up the reputation of all parties involved.

Lessons learned

This case study has provided useful insights into enhancing alliancing performance of a construction project in that both appropriate construction and organizational culture are vital to implement alliance approach, process and spirit successfully.

Case study 10

National Museum of Australia in Canberra, Australia – A case study of a large-scale public sector project.

Background

This was a large-scale public sector project with the contract sum of approximately AU$150 million. The project began in 1998 and was completed in 2000. There were about 300 participants involved in this project: 10 from the client, 6 from the main contractor, 15 from the consultants, and 260 from the subcontractors. The client was the National Museum, the main contractor was Bovis Lend Lease, and the main consultant was Peck Von Hartel (architect).

Alliancing approach and process

At first, the client worked with the consultant and explained to the members of the Australian Parliament that if the traditional form of contract was adopted in this project, it would lead to adversarial relationship between parties, thus probably attracting many claims and disputes. They then persuaded the members of the Australian Parliament to adopt alliance contract. Although this was a tough job, the members of the Australian Parliament finally agreed to use this innovative approach to procure this project.

Interview results

Major reasons for adopting an alliance contract

The major reason for adopting an alliance contract in this project was that the client, as persuaded by the chairman, believed that it was impossible to use the traditional form of contract to manage this project, without leading to adversarial relationship amongst different parties.

Perceived major benefits and critical success factors for adopting an alliance contract

The perceived major benefits were that the project was completed on time, within budget and to a high quality. The CSFs for adopting an alliance contract in this project, as stated by the interviewee, could be summarized as a short sentence: 'Parties behaved how they should behave'. In other words, the client selected the right team.

Relationships in the alliance contract

The working relationship between the client and the project team was very good because when problems arose, parties were willing to follow the alliance contract.

Major difficulties in alliancing implementation

A number of technical difficulties arose in this project but they were resolved smoothly by an integrated project team because they worked together to solve the problems.

Effect of construction culture on alliancing practice and performance

The interviewee considered that the construction culture in Australia was still very adversarial and the unions were very strong and this culture was unhelpful to alliance practice and performance in this project.

Effect of organizational culture on alliancing practice and performance

The interviewee described his company's culture as 'excellent'. However, he stated that since he was only a chairman in this project, the culture in his organization did not affect the alliance approach and performance in this project.

Lessons learned

This was the first construction project adopting alliancing approach. Therefore, it provided valuable insights into formal procedures for implementing successful alliance contract in construction. It acts as an exemplar all over the world.

Case study 11

Tullamarine Calder Interchange project, Australia – A case study of an infrastructure project.

Background

This was a road infrastructure project that began in January 2005 and was completed in 2007. The contract sum was approximately AU$160 million and there were four major parties, including a client, two main contractors and a consultant, and they formed a consortium. The client was Vic Road, the main contractors were Parson Brinkhowf Ltd and Baulderstone Ltd, and a main consultant was Currie & Brown (independent expert).

Alliancing approach and process

At first, the client initiated an alliance contract in this project. Then, he recruited an independent alliance facilitator to assist him to employ the most capable project team through drafting tender documents, and conducting interviews and alliance workshops. Four project teams submitted their proposals and they were assessed through interviews and workshops. Finally, the most capable team won the project. After that, a consultant was employed (actually, it was an uncommon practice). Having gone through the design stage for six months, all the parties drafted the alliance contracts together (a key component was TOC) and they agreed the schedule, budget, and other project details.

Interview results

Major reasons for adopting an alliance contract

The interviewee viewed that there were four major reasons for adopting an alliance contract in this project, encompassing: (1) resources were limited; (2) the schedule was very tight; (3) the client considered that alliancing was value for money; and (4) the project was complex and large. It was of interest to note that this was the first project for the Victoria government to use an alliance contract.

Perceived major benefits and critical success factors for adopting an alliance contract

The interviewee opined that the perceived major benefits for using an alliance contract were: (1) building up good working relationship; (2) building up good reputation; (3) building up long-term business opportunities; (4) improving the way that they worked together; (5) reasonable profit; and (6) training.

The CSFs for adopting an alliance contract in this project included: (1) the capable team; (2) good working relationship; (3) mutual trust; and (4) parties were willing to share resources.

Relationships in an alliance contract

The working relationship between the client and the project team was excellent, for instance, they were willing to share offices.

Major difficulties in alliancing implementation

The interviewee mentioned that it was not suitable for him to comment this issue.

Effect of construction culture on alliancing practice and performance

The interviewee perceived that people worked more collaborative than before and they were more focused on health and safety, and it is a growing industry when compared with the UK. This culture was very helpful to alliance practice in this project because good cooperation and collaboration was a fundamental element for successful alliancing. Since there were some

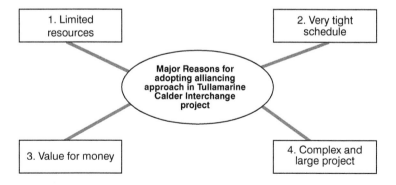

Figure 12.9 Major reasons for adopting alliancing approach in Tullamarine Calder Interchange project

agreed KPIs for the project team to achieve, these KPIs acted as strong drivers for the project team.

Effect of organizational culture on alliancing practice and performance

The interviewee mentioned that the organizational culture in his company was good in that colleagues discussed issues openly and solved problems together with high efficiency. This culture had no effect on selecting contractors but affected much on the negotiation of TOC.

Lessons learned

The lessons learned from this successful alliancing project could help the practitioners to minimize the harms brought about by potential difficulties and maximize the benefits gained from implementing alliancing concept. It is confirmed that alliancing approach is especially suitable for those construction projects that are large and complex in nature, with high risks and great uncertainties.

Case study 12

Lawrence Hargrave Drive Alliance project, Australia – A case study of an infrastructure project.

Background

This was a road infrastructure project and it began in November 2003 and was completed in December 2005. The contract sum was approximately AU$50 million and there were four major parties (a client, a main contractor and two consultants) involved in this project, with approximately 20 management staff and 100 on-site workers. The client was the Roads and Traffic Authority (RTA), the main contractor was Barclay Mowlem Construction Ltd, and the two main consultants were Maunsell Ltd and Coffey Ltd respectively.

Alliancing approach and process

At first, RTA initiated to use an alliance contract in this project and called for interest by invitation. Then, RTA recruited an independent alliance facilitator to assist them to employ the most capable project team through drafting tender documents and conducting interviews and alliance workshops. Three to four project teams submitted their proposals and they were assessed through interviews and workshops. Finally, the most capable team won the project. Afterwards, the project team (including the client) implemented the alliance contract with three phases. The first phase was to develop a best solution (70 possible solutions were brainstormed to solve complex problems). The second phase was to prepare the TOC budget, and the third phase was to construct the project.

Interview results

Major reasons for adopting an alliance contract

The major reason for adopting an alliance contract in this project was that there were a lot of uncertainties and risks for RTA (it was unstable because it was unsafe for the community: serious landscape problems and rock falls). Therefore, to shorten the timing to complete this project, RTA would like to use fast-track method to tackle these problems.

Perceived major benefits and critical success factors for adopting an alliance contract

There were seven perceived major benefits when an alliance contract was adopted. These included: (1) more flexibility; (2) more openness; (3) more efficient decision making; (4) faster reaction and quicker problem solving; (5) faster project delivery; (6) minimizing cost; and (7) excellent quality. And the CSF was mainly because the mechanism of alliance contract itself made this project successful.

Relationships in an alliance contract

The working relationship between the client and the project team was good because most of the project team members worked in the same direction (good cooperation and collaboration with no competition).

Major difficulties in alliancing implementation

Although the majority of the project team members integrated quite well, some of them had individual goals (this problem was not solved). On the other hand, there was limited space to conduct the works (most of the problems were resolved but with much effort, however, some of them could not be tackled).

Effect of construction culture on alliance practice and performance

The interviewee reckoned that the construction culture in non-residential buildings (such as shopping centres, warehouse, apartments, hospitals and

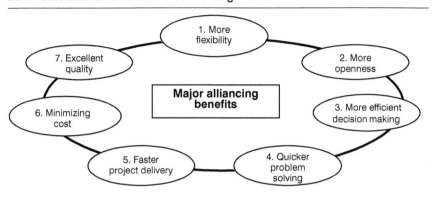

Figure 12.10 Major alliancing benefits for Lawrence Hargrave Drive alliance project

other high-rise buildings) was very confrontational because people were commercial-driven. However, the construction culture in civil engineering works was more cooperative because people focused more on project delivery and quality. The construction culture in the mining and resources sector was much more cooperative because people focused much more on safety, quality and project delivery (timing is crucial).

The interviewee perceived that it was unlikely to use an alliance contract in building works under the Australian construction culture. However, an alliance contract was suitable to be implemented in civil engineering works because parties tended to be more cooperative (in other words, this culture was helpful to adopt an alliance practice in this project.)

Effect of organizational culture on alliancing practice and performance

The interviewee described that the size of his previous company (Barclay Mowlem Construction Ltd) was very large and there were different kinds of people in his company (all types of personality). Some of them were suitable for alliance contract because they were more flexible. However, some of them were too 'single-focused' (too money-minded). Since only those capable people who suited alliance contract well were selected to participate in this project, they helped this project much due to their competence and flexibility.

Lessons learned

This case study has supported an assertion that an alliance contract is especially well suited to those construction projects that are complicated in nature, with high risks and great uncertainties inherent in the project. It is recommended that project alliancing, when properly implemented, can generate a good workable model for different parties to work more effectively and efficiently, thus ultimately achieving win-win outcomes for all parties involved.

Comparisons of the case studies in alliancing between the UK and Australia

Similarities

Many selected case studies in alliancing in the UK and Australia indicated that selection of appropriate partners is a key to alliance success for both construction and non-construction projects. For non-construction projects, partners should be selected when they have the same business growth objectives, strategies and long-range visions for which the new business operates. For construction projects, partners, especially contractors, should be selected based on their own capability, professional attitude and resource management, but not based on price only.

It has also been found that an alliancing approach in construction is particularly suitable for projects that are large and complicated in nature, facing tight schedules with high risks and great uncertainties inherent in the project. It is highly recommended that an alliancing approach, when implemented appropriately, could generate a good practical model for different partners to work more effectively and efficiently, thus eliminating potential project risks and enhancing certainties in implementing the project.

Differences and highlights

Based on the aforesaid case studies in alliancing, it can be concluded that many alliancing projects were implemented successfully in the UK and Australia. However, the reasons for their success are not completely the same and they deserve to be highlighted.

Alliancing approach suitable for small-scale projects

A case study in alliancing from Australia showed that in addition to large-scale projects, alliancing approach is also suitable for small-scale construction projects if the projects are complex in nature, with high risk and great uncertainty inherent in them.

Appointment of a capable external facilitator

A case study in construction alliancing revealed that the appointment of a capable external facilitator may be useful for alliance success because the competent external facilitator may be able to assist the client to draft well-suited alliance contract and select the most capable project team.

Alliance approach suitable for good construction and organizational culture

Some case studies in construction alliancing showed that good construction and organizational culture are crucial to implement alliance approach, process and spirit successfully and vice versa.

Chapter 13

Case studies in public–private partnerships

This chapter shows how PPP are implemented in two different nations. A total of six case studies in PPP in the UK and Australia are selected for demonstration purposes.

Case study I

Project 2002 – Enhancing the Quality of Education in Glasgow City Schools, Scotland, UK, by PPP (extracted from the website of Hong Kong Efficiency Unit http://www.eu.gov.hk/english/case/case.html) (permission has been obtained for both print and online use from the Hong Kong Efficiency Unit).

Background

There were a total of 39 secondary schools in Glasgow in 1996. A majority of them were poorly conditioned, and they needed substantial refurbishment with accumulative maintenance costs estimated to be more than £100m. In spite of the fact that the schools could accommodate more than 50,000 pupils, there were only about 29,000 pupils in the city. It was expected that this figure would rise slightly in the coming future. The sponsor for the project was Glasgow City Council, Scotland. The consortium was 3Ed (Miller Group Limited, Amey Ventures Ltd and Halifax Projects Investments Ltd). The capital cost was £225m, with the contract length of 30 years and information communications technology (ICT) of 12 years.

To comply with modern standards, there was an urgent investment (approximately £150m) for the schools to spend by August 2002. Nevertheless, the costs of the enormous surplus in accommodation cut into the funds available for this purpose. In the face of this challenge, the sponsor, Glasgow City Council, decided to close 10 schools, but it still provided for at least 32,500 pupil places in 29 strategically placed secondary schools. The quality of accommodation of these schools was good. They could provide a high-quality learning environment with first class ICT. The Glasgow City Council, however, could not be able to commit to the essential level of investment from within its own resources. It sought help with the Scottish Office/Scottish Executive and the biggest PPP project in the UK education sector was thus created.

Aims and phases of project 2002

The aims of Glasgow City Council's Project 2002 were to ensure that all schools be able to: (1) support the development of basic skills with help from new technology; (2) develop core ICT skills for all young people; (3) access the best current information to support resource-based/self-directed learning; and (4) support the full implementation of the 5-14 programmes in the first and second years along with other vital Council and Scottish Executive initiatives. There are three phases in the contract (Table 13.1).

Project design and partner selection process

The directors of the education and financial services submitted a joint report on a feasibility study for a PPP project in education in 1997. In February 1998, Glasgow City Council decided to decrease its secondary school estate from 39 to 29 secondary schools. In addition, it intended to reinvest the savings and prepare a business case to assess the affordability and value for money of a PPP approach for the modernization of its secondary schools. After going through extensive negotiation and consultation with the bidding contractors, the Council finally selected a preferred bidder together with a first reserve.

Based on the detailed output specification of Council for the accommodation services required, each bidding contractor produced a

Table 13.1 Phases and Tasks of Project 2002, Glasgow (reproduced from the Hong Kong Efficiency Unit) (http://www.eu.gov.hk/english/case/case.html) (with permission for both print and online use from the Hong Kong Efficiency Unit)

Time Frame	Tasks
Phase 1: April 2000–August 2002	The contractor has to ensure that all 29 schools meet modern standards. This involves building new schools, extensions, and the refurbishment of all retained buildings.
Phase 2: August 2002–August 2029	The contractor has to maintain all 29 schools to a high standard through regular planned maintenance to each building and prompt resolution of building problems/issues. Planned maintenance costs are estimated to be about £200m in total over this period.
Phase 3: August 2029	The contract ends with the city council resuming responsibility for building maintenance. There will be no hand-back costs to the council and buildings have to be free of any major maintenance problems.

solution. Bids were then compared during the extensive tendering process. They were also compared against the public sector comparator (PSC), which is the estimated cost of meeting the same output and user requirements demanded of the private partners but through a publicly funded solution. The bid of 3Ed consortium was evaluated as providing best value. It was thus selected as the Council's preferred partner for Project 2002.

During the contract, the Council owns the buildings and the contractor is given a lease or licence to operate in the buildings. The contractor is responsible for building and grounds maintenance, security, cleaning, heating, maintenance of fabric and fittings and the planning, and development and management of ICT.

Advantages of public-private partnerships over traditional project management

There were three major advantages of adopting PPP approach in this project, including innovative financing, risk transfer and change management.

Innovative financing

It is reported by Hong Kong Efficiency Unit that without Project 2002, Glasgow City Council would have had to close the 10 surplus schools, invest in the refurbishment and upgrade school facilities on its own. But instead of using conventional public funds, the private partner, 3Ed, invested £220m over two years to refurbish existing schools, build extensions and replace some old schools with new ones. 3Ed also invested over £15m for the ICT development.

Besides, reallocation of the savings made from the rationalization programme has enabled further improvements, encompassing raising staff levels, improving resources (including ICT) and improving school accommodation. To ensure that the Council can afford the project, the Scottish Executive provides a huge amount of extra resources each year.

Transfer of risk

A core element of the PPP project is to transfer risks from the Council to the contractor. For instance, the provision of accommodation has to be fit for purpose and reflect wholly the output specifications of Council. If there are flaws in the solution, the contractor will experience harsh financial penalties and will have to remedy the problem at its own expense.

Change management

Based on an understanding of education processes, Project 2002 aims to benefit teachers by engaging them more effectively in the learning process. For instance, the project enables teachers to use ICT in the development of learning and assessment of materials. In addition, they can access the most current learning resources via the internet, develop proficiency in the core ICT skills and greater confidence in adopting both traditional and newly developing teaching techniques. The ICT contract provides for a fully managed service with the contractor and the contractor is responsible for ascertaining that the equipment, software and networks work effectively and efficiently. Schools can still change the exact make-up of their ICT provision, which depends on their individual requirements.

Project 2002 not only satisfies current needs, but it also aims to provide accommodation and ICT services that are adequately flexible to facilitate the changes of the future. In addition, it makes access to ICT available to all pupils and teachers to enable them to work productively and efficiently together. By doing so, it helps to raise standards and maximize the individual potential of every participant.

Lessons learned

The lessons learned from this project encompass:

1 Testing of value for money: the public sector can get money more cheaply than the private sector can borrow. But the private sector can provide better value for money by more effective management of the risks related to procurement and continuous operations, thereby offsetting any higher financial costs. Best value for money can be achieved through the optimal combination of whole life costs and benefits. In general, three separate value for money decisions have to be taken, including whether to proceed with the project at all, whether to proceed using PPP and selecting the best private sector partner.
2 Achieving affordability: though PPP could help to transfer risks, the buildings still have to meet the high standards set and a good design has to be secured within the constraints of affordability.
3 Consultation: It has been proved during the partner selection process that regular consultation with school boards, head teachers and the PPP coordinators of schools facilitated better design and acceptance.

Case study 2

University College London Hospital (UCLH) redevelopment – Improving the Standard of Healthcare by PPP (Extracted from the website of Hong Kong

Efficiency Unit http://www.eu.gov.hk/english/case/case.html) (permission has been obtained for both print and online use from the Hong Kong Efficiency Unit).

Background

The National Health Service (NHS) in the UK encounters many problems arising from its stock of outdated buildings and services. The costs of renovation and new development are high. To satisfy changing needs and demands, the government encourages health service trusts (the local health providers) to take more responsibility to provide facilities and support services to specialist private sector managers. To do so, it allows its own staff to focus on providing high-quality health care. The adoption of a PPP approach in this project proves to be effective in improving healthcare by (1) promoting private investment in the capital assets and operational management of hospitals; (2) improving value for money by allocating risks to parties that are best able to manage them in the public or private sectors, thus providing higher incentives for sustained and effective performance improvement; and (3) upgrading and rationalizing health service authority property.

An important element of PPP in health care is that the NHS defines its needs in terms of 'output'. The output here refers to the nature and level of the service required. It invites private sector bidders to present their solutions to meet these service needs. This permits the private sector to make the fullest possible use of its experience and skills so as to bring novel solutions to the needs of the health service.

The NHS Trust at UCLH was composed of eight hospitals, including: (1) University College Hospital; (2) Eastman Dental Hospital; (3) National Hospital for Neurology and Neurosurgery; (4) the Middlesex Hospital; (5) Hospital for Tropical Diseases; (6) Elizabeth Garrett Anderson and Obstetric Hospital; (7) the Heart Hospital; and (8) the Royal London Homoeopathic Hospital. It spreads across central London and is one of the UK's largest providers of healthcare services, medical research and training. Nevertheless, many of the facilities that provide clinical care and services are housed in out-dated, rigid and confined buildings.

Patients and staff must shuttle back and forth between hospitals for tests and treatment. Due to the dispersion of clinics, wards and operating theatres on various sites, it resulted in unnecessary duplication of facilities, equipment and support services costing up to £12m a year. In 1995, a major redevelopment scheme was proposed to construct a new hospital and development of new sites for existing services. By doing so, it can help to replace the old, scattered buildings and to modernize the services. A huge amount of investments were also planned for acquiring new equipment and ITC systems.

However, the UCLH Trust could not be able to commit to the indispensable level of investment on its own. To solve this problem, it undertook detailed evaluation of the risks, benefits and value for money of different funding options, encompassing (1) the 'minimum only' option (the minimum only option means using minimal development to sustain facilities which could deliver the core services of UCLH, meet requirements of clinical practice, maintain accreditation and keep up with the charter of patients); (2) the conventional funding option; and (3) the PPP option. After the evaluation, it was proved that the PPP option was the least costly approach and offered the best value for money. Thus the PPP scheme in UCLH was adopted in which it was the largest and most complex PPP project in the UK healthcare sector.

Aims and project design of the UCLH redevelopment scheme

The aim of the UCLH redevelopment scheme is to modernize the services of the trust so as to meet changing demands and to improve upon the standard of care provided to patients. To achieve this target, most of the services that are provided at three of eight scattered sites will be centralized into a new 669 bed acute hospital in central London. The construction started in November 2000 and the main hospital was planned for completion in April 2005 with the remaining facilities due in June 2008.

The UCLH scheme is a typical design, build, finance and operate (DBFO) scheme. And the private sector partner, the Health Management (UCLH) Plc consortium, is responsible for (1) designing the facilities; (2) building the facilities (to time and at a fixed cost); (3) financing the capital cost; and (4) operating the facilities. During the contract, the trust will utilize the building and the non-clinical services provided. The contract terms detail the services range and the performance standards against which the services will be monitored. The trust makes no payments until services are provided to the consented standard. A penalty point system will be implemented and the performance standard has to be strictly maintained to ascertain the subsequent full payment. After the agreement is expired, the building in agreed condition will be returned to the trust. If further capital investment is required in future, the initial contract will be changed to provide for the manner in which this should be managed.

The trust is still the hospital owner and remains in charge of providing high quality clinical care to NHS patients. Although it goes on to be the employer of the NHS clinical staff, around 450 non-clinical staff will be transferred to work for the facilities management company. The trust, the unions and the consortium have designed a change management programme to conduct consultation and provide counselling and support down to individual staff level. No changes to terms and conditions of employees will be made unless they are previously agreed with the trade unions. Any surplus staff issues stemming from efficiency improvement will be addressed through natural wastage.

Partner selection process

The UCLH project followed the NHS PPP process and required the trust to (1) set up the strategic context, evaluate the options and to make the case for alteration in a strategic outline case before obtaining approval; (2) identify and develop a favoured choice through an investment appraisal, and make the case in an outline business case; (3) prepare for procurement by changing the approved option into a detailed output and outcome specification, and desired risk allocation between the trust and the successful bidder; (4) advertise the project, identify potential bidders and the best privately financed solution; (5) choose the best bidder with whom negotiations could be completed, involving stakeholders in the proposals' assessment; (6) complete the definitive investment appraisal and full business case to get final approval from HM Treasury; (7) finalize, award and implement the contract; and (8) assess and monitor the project.

Like Case Study 1, there were three major advantages of adopting PPP approach in this project, including innovative financing, risk transfer and change management.

Lessons learned

The lessons learned from this case study encompass:

1 Value for money testing: though it is less expensive for the public sector to obtain money than the private sector, the private sector can compensate for any higher financial costs by providing better value for money. This can be done through its commercial insight, innovation and more effective risk management related to procurement and continuous operations. Best value for money is achieved through the optimum combination of whole life costs and benefits. In general, commissioning bodies must take three separate value for money decisions, including (1) whether to proceed with the project at all; (2) whether to proceed using PPP approach; and (3) selecting the best private sector partner.

2 Achieving affordability: though many of the risks are allocated to the contract, the public sector still need to ensure that the desired high standards and good designs can be secured within the affordability constraints.

3 Inclusion of property: the NHS did not formerly own the site for the new hospital and the trust must acquire the freehold and leasehold interests of sites before the financial close. In addition, a collaborative

approach to risk sharing, particularly during the construction phase, was required.

4 Consultation: in the redevelopment of UCLH, better design and acceptance was facilitated when the public was consulted after the publication of the outline business case. It was necessary to work with patients, staff, the local community, and other health providers to plan the new UCLH. Consultation not only assisted with tapping views on how to ensure the adaptability for potential changes in clinical practices, but it also helped to forecast interim services and transitional arrangements during the relatively long construction period of eight years.

Case study 3

Project Ministry of Defence Estate in London (extracted from the website of Partnerships UK: http://www.partnershipsuk.org.uk/PUK-Case-Studies. aspx) (permission has been obtained for both print and online use from Partnerships UK).

Background

Defence Estates has started a novel partnership with the private sector in which the locked-in value of its estate was adopted to provide new facilities at a core site within a tightly managed programme. The principles of the innovative Project Ministry of Defence Estate in London (MoDEL) delivery programme can be replicated across many other areas of the public sector estate. Co-sponsored with the Defence Estates in the procurement, Partnerships UK has produced a summary to assist public authorities in understanding MoDEL and considering whether a similar approach is suitable for them.

The project stemmed from a strategic review of the MoDEL. It involves new investment in a major site at MoD's West London Airfield at Northolt. In addition, personnel are relocated from six sites elsewhere in London and the surplus land is subsequently disposed. The rationalization project will deliver both financial and operational benefits to MoD, encompassing the accommodation upgrading, improvement in operational capability and efficiency savings from moving various sites onto a single core location.

Transaction summary

Defence Estates (representing Ministry of Defence) has entered into a contract with VSM Estates. VSM Estates is a consortium between Vinci plc and St Modwen Plc selected in competition. With VSM managing and financing the programme, the transaction comprises three key elements, including (1)

VSM will handle and finance a programme of new investment (*c.* £150m) with fixed price in the core site at Northolt; (2) VSM will relocate some of 900 personnel from six other sites and manage the subsequent disposal of the surplus land over a 5–6 year period; and (3) VSM has underwritten minimum land values (*c.* £230m) and will invest its resources in achieving planning permission at the six surplus sites, the demolition of existing buildings and installing basic infrastructure before handling their disposal as packages for development on the open market.

Commercial principles

The underwritten land values are adequate to cover the delivery of the investment works at Northolt. These encompassed the costs of interim finance during the programme. Sales receipts above the underwritten land values will be shared between VSM and Defence Estates.

The equity returns of VSM are earned completely from the sale of surplus land above the underwritten values and it is thus incentivized to maximize receipts and deliver the programme on time. Any extra costs resulting from alterations in requirement and/or delay are paid by Defence Estates potentially out of its share of land value receipts and it is thus incentivized to take well-timed decisions.

Underpinning the contract are major commercial principles and terms that line up the interests of the public sector with the contractor:

1 The surplus sites are sold immediately, with MoD taking a leaseback for the continuing occupation period to provide VSM with adequate interest upon which to raise finance;
2 By using public procurement, VSM has to subcontract any extra works required out of the initial specification, thus demonstrating best value and preventing from the creation of a monopoly supplier;
3 VSM cannot carry out direct development of surplus sites (other than investment infrastructure), so it will offer packages for sale on the open market to demonstrate maximum sales receipts; and
4 MoD is indebted to work with VSM to deliver the agreed programme and to collaborate in moving surplus sites.

Procurement

Defence Estates considered procuring the project through traditional means (i.e. a separate works contract and individual site sales) and a combined PFI contract. It selected the novel 'Prime Plus' contract as the best means of securing the objectives of projects, by combining the pricing guarantees of a traditional approach with the programme management and financial disciplines of a PFI. Following a period of original preparation, the procurement began in August 2004 and the contract was awarded on time and within budget in August 2006 after going through a strong competition.

Value for money

The benefits for MoD are clear and include (1) an accelerated programme; (2) realization of operational efficiency savings; and (3) sharing in the returns from private sector infrastructure investment and planning skills. As the private sector partner, VSM is appropriately incentivized to manage the programme delivery faster and more efficiently than MoD acting alone. It is able to understand and manage the risks involved, and it is capable of investing specialist resources. Financially, VSM is taking key risks in relation to the pricing of the works at Northolt, the levels of receipt that can be achieved on land sales in the future and its ability to manage the programme to minimize the financial costs related to negative cash flow.

Lessons learned/opportunities for replication

The approach adopted in this case study can be considered in circumstances where there is a well-articulated estate rationalization plan and where one or more of the following features exist: (1) there is a requirement for investment in new facilities; (2) focus on a core facility will permit peripheral locations to become surplus over time; (3) the properties concerned are preferably within a single ownership and principally freehold or valuable leasehold; (4) there are benefits from integrating the programme to deliver new investment and associated rationalization. It is unnecessary for all the properties to be in the same region or local authority area, nor it is essential for the value of surplus properties to exceed the new investment costs.

Case study 4

Provision of Victoria County Court Complex, Australia by PPP (extracted from the website of Hong Kong Efficiency Unit http://www.eu.gov.hk/english/case/case.html) (permission has been obtained for both print and online use from the Hong Kong Efficiency Unit).

Background

The County Court was spread across four different locations in Melbourne. The old buildings were badly maintained; there were inadequate courtrooms, or space in the courtrooms, to meet the increasing demands on the facilities. As a result, backlogs in both the criminal and civil areas happened. There was also in lack of basic services, such as disabled access, and insufficient areas for jurors, witnesses, victims and the public. In December 1999, the state government decided to house the County Court in one location and the government was prepared to work with the private sector to provide the new justice infrastructure and non-core services. However, the essential court services would continue to be provided by the government itself.

In June 2000, a private sector consortium, The Liberty Group Consortium Pty Ltd, was contracted under a Court Services Agreement to design, develop, build, finance and maintain a new County Court complex in Melbourne. It is the largest court complex in Australia. The commissioning body is Department of Justice, Victoria, Australia, with a total of capital cost of AU$140 million and a contract length of 20 years. The contractor would provide the building related services, such as building security, maintenance and ITC support. The government would provide the judicial related services, such as the justice administration, management of case list, registry, court information, recording, reporting, transcription of trials, and contract supervision and administration.

Private sector consortium selection process

The government provides community input into development of infrastructure through suitable planning mechanisms. Where there is private sector involvement in major public infrastructure projects, the most suitable contractor will be selected through a rigorous and transparent public tendering system. Before formal consideration of a PPP scheme, the Department of Justice submitted a project proposal to the Department of Treasury and Finance. It then reviewed the conformity of the proposal to government policies, its viability and its priority. Approval by the Victoria Cabinet is required for major infrastructure projects; it occurs before expressions of interest from private sector consortia are sought and before the release of a project brief to short-listed bidders. The release of the project brief is a major milestone because it signals that the government is prepared to proceed with the project once a conforming bid offering value for money in comparison with the public sector comparator is received.

Like Case Studies 1 and 2, there were three major advantages of adopting PPP approach in this project, including innovative financing, risk transfer/ allocation and change management.

Lessons learned

The lessons learned from this case study include:

1 Combination of separate court buildings and integration of non-core services: the private sector can provide both court complexes with good quality and comprehensive non-core services in an integrated and cost-effective way. Non-core facilities well-suited with a court

facility can be considered within a complex. It can help to facilitate the legal profession and the administration to compensate for the cost of providing new facilities.

2 Change management: the private sector can be contracted to handle the migration from old facilities to a new complex.

3 Evaluation of service delivery options: an economic appraisal of different internal and external service provision options should be undertaken so as to enable the most cost-effective outcome and an appropriate assessment of private sector tenders.

4 Specification of service requirements and service levels: a court services agreement ought to clearly define the accommodation and non-core court services to be provided.

5 Performance review and monitoring: the contract administrator ought to assess the service delivery performance of the contractor against the outcome specifications in the contract.

6 It is important to ensure staffing quality in service delivery.

7 Contract arrangements for dispute resolution: it should clearly specify the form of default notices, the rights and obligations of the respective parties and the dispute resolution process in an agreement.

8 Continuous technology improvement: arrangement should be considered whereby the contractor earmarks essential resources to acquire and install up-to-date equipment and products, or upgrade existing products to meet evolving requirements for ITC services.

9 Supply of source code and intellectual property for ITC services: the release of up-to-date source code for software products should be specified in agreement. In addition, it should specify to grant a licence to the government or its authorized persons to use all the intellectual property rights in the IT services.

Case study 5

Water Treatment Plants by PPP – Central Highlands Region Water Authority, Victoria, Australia (extracted from the website of Hong Kong Efficiency Unit http://www.eu.gov.hk/english/case/case.html) (permission has been obtained for both print and online use from the Hong Kong Efficiency Unit).

Background

The Central Highlands Region Water Authority (the Water Authority) was established in July 1994. It is wholly a state government-owned business enterprise. As one of the 15 non-metropolitan urban water authorities in Victoria, Australia, the Water Authority is responsible for providing water and wastewater services. The services range from treatment and delivery of drinking water to customers' homes and the collection, treatment and disposal of sewerage and trade effluent. The Central Highlands Region is centred on Ballarat and encompasses a number of towns in the vicinity. The Water Authority has been busy because it extended Ballarat's water supply to nearby towns where water quality did not meet Australian Drinking Water Guidelines. Towns in the vicinity relied on old infrastructure that was incapable of delivering high-quality water to customers.

The Water Authority identified a potential opportunity for private sector to participate in providing substantial water quality improvements without extra charges on customers. In April 1999, the Water Authority signed a build, own, operate and transfer (BOOT) contract with the winning consortium, Thames Water Ballarat Pty Ltd and United Water, to supply Ballarat and nearby towns with treated water for a 25-year period. This was one of the first non-metropolitan water BOOT projects in Victoria. In November 2000, United Water accomplished construction of two new water filtration plants on land. The two plants were owned or leased by the Water Authority at Ballarat's two largest reservoirs. The treatment plants are each sized to produce 65 million litres of treated water per day. Both the new treatment plants and the existing treatment and pumping facilities are managed, operated and maintained by United Water. A significant improvement in water quality has been seen, for example (1) the bacteriological compliance results jumped from 67% in 1999–2000 to 82% in 2000–2001; and (2) the proportion of samples that met Australian Drinking Water Guidelines: colour was 91% in 2000–2001 (82% in 1999–2000) and pH 90% in 2000–2001 (84% in 1999–2000).

Partner selection process

The partner selection process for this project is the same as Case Study 4.

Advantages of PPP over traditional project management

There were three major advantages of adopting PPP approach in this project, including innovative financing, risk allocation and change management.

Innovative financing

A major advantage of PPP is to promote private investment in the capital assets, management, operation and maintenance of water treatment facilities. Under PPP, services not assets are purchased. United Water supplies Ballarat with treated water for a 25-year period. During the contract period, the Water Authority is obligated to pay United Water a yearly service charge comprised of both fixed and variable components (AU$4.7 million in 2000–2001). At the end of the contract, United Water will hand over the ownership of the two water treatment plants and related facilities back to the Water Authority.

Risk allocation

Another advantage of using PPP is to improve value for money by allocating risks to those parties best able to manage them, and providing incentives for sustained and effective performance improvements. As mentioned previously, the fundamental principle in PPP risk allocation is that each individual risk is identified and then allocated to the party best able to handle that risk. In this project, most of the risks associated with the building, operation and maintenance of the new water treatment plans are allocated to the private partner.

Change management

The project offered the Water Authority an opportunity to identify and address the needs of community relating to water supply and, in particular, problems of poor water quality in towns that relied on old water supply infrastructure. The project also provided a chance to predict the future community needs with taking into consideration of changes in population and aspirations of residents in the region, in the light of Australian and international water quality guidelines. As part of the contractual arrangements, United Water is required to transfer its skills and knowledge to the Water Authority, which has consequently improved water quality to customers in different water distribution zones.

Lessons learned

The lessons learned from this project encompass:

1 Well-established practices: the Ballarat project is only one example of many successful PPP projects adopted to improve the quality and cost-effectiveness of water supply services.
2 Evaluation of service delivery choices: to enable the most cost-effective outcome and a correct assessment of private sector tenders, an economic appraisal (cost and benefit analysis) of the different internal and external service provision options should be undertaken.
3 Specification of service requirements and service levels: a water supply service contract should define very clearly the services to be provided.
4 Day-to-day operation and maintenance of water supply infrastructure services can be delivered by the private sector in an efficient and cost-effective way.

Case study 6

Provision of Vehicle Fleet Services, Victoria, Australia, by PPP (extracted from the website of Hong Kong Efficiency Unit http://www.eu.gov.hk/english/case/case.html) (permission has been obtained for both print and online use from the Hong Kong Efficiency Unit).

Background

The Victoria Auditor-General recommended improvements to the Victoria government motor vehicle fleet in May 1993. The motor vehicle fleet was composed of 10,700 vehicles and it was considered to be poorly supervised and under-utilized. The state government subsequently reduced the fleet size by 15% and generated sale proceeds of about AU$12 million in June 1993. This resulted in annual savings of AU$6 million. The government also sought to achieve additional savings of AU$15 million each year through fleet rationalization and better management. A private company, JMJ, was contracted to deal with 3000 vehicles on a trial basis for one year starting from July 1993.

Sale and lease-back of government vehicles

The government examined the available leasing options for its motor vehicle fleet in May 1994. It aimed to facilitate improvements in fleet management, generate a huge amount of cost savings and reduce interest costs by applying the sale proceeds towards the state debt retirement. The corporate advisers in the Department of Treasury and Finance (DTF) in Oct 1994 considered that a private sector operating lease arrangement had the potential to deliver benefits effectively. It was expected that savings of AU$139 million over a period of 10 years could be achieved. In 1995, the DTF called for expressions of interest (mainly banking, financial institutions and fleet operators) for the submission of proposals to provide fleet financing and fleet management services of the existing fleet of around 8,500 passenger and light commercial vehicles.

The results of the tender exercises were:

1 The DTF selected the Commonwealth Bank of Australia (CBA) in September 1996 to be a fleet financier for the sale and lease-back of the government's vehicle fleet under a seven-year master lease agreement; and

2 Both LeasePlan Australia Ltd and GE Capital Fleet Services (now known as GE Fleet Services, a division of the US company General Electric) were appointed as external fleet managers in June 1997 to provide fleet management services for a two-year period and renewable at the

government's option. The DTF maintained its overall responsibility for managing these external contracts.

The DTF estimated the sale proceeds of 8,500 vehicles at AU$190 million in September 1996. It would be used to repay state debt and could result in interest savings of AU$14 million per year. Further savings from the sale and lease-back arrangements were about AU$30 million per year.

Agreement outline

Master lease agreement

After reviewing vehicle utilization, the government decreased the fleet size by 15% to 7,200 in July 1997. These were sold to the CBA for AU$168 million and leased back. As at June 2002, the annual rental payments were about AU$40 million for 7,951 vehicles (with a leased value of around AU$219 million). There were some key terms in the master lease agreement. These included:

1 The government had the right to terminate the financing facility at three years, six years or annually after seven years. The CBA could not terminate the financing facility before year seven.
2 The terms of the individual vehicle depended on the particular vehicle and usage thereof. In broad terms, the rental term would expire when the vehicle had reached 40,000 km or two years' service from the date of purchase. Individual vehicle leases might be ceased if the government did not need them.
3 The government was required to make monthly lease payments in arrears in relation to each vehicle. Payments were primarily based on the vehicle purchase price, any prepaid services charges and the cost of any vehicle accessories or optional equipment.
4 The state was required to fund any deficit between the value of vehicles sold at the end of the lease term and the agreed residual value of the vehicles via increased rentals on new leases, or if there were a profit, the state would obtain the benefit at the expiry of the lease facility.

External fleet managers

The master lease agreement permitted fleet management services to be provided by either private sector providers or government agencies. The services of external fleet managers (LeasePlan and GE Fleet Services) included (1) purchase of new and replacement vehicles and connection with the financier; (2) administration of vehicle registration and insurance; (3) collection and reporting of asset management and vehicle operating data;

(4) coordination of servicing, maintenance and repair of vehicles; and (5) disposal of vehicles through auction houses.

A fleet which was composed of about 7,600 vehicles was managed by external and internal providers. The distribution was as follows: (a) external: LeasePlan (23%); (b) external: GE Fleet Services (30%); and (c) internal: other government agencies (47%).

Staffing issue

Except for the vehicles used by ministers, the Victoria public service practice often involves a vehicle user driving the vehicle himself or herself (user-driver). As a result, the project did not involve any considerable staff redundancy/transfer issues of vehicle drivers.

Other developments

LeasePlan submitted an offer to the DTF to renew the contract at a revised fee prior to the expiry of its contract in May 2000. However, it suddenly ceased negotiations with the DTF. As a result, the DTF set up an in-house fleet management unit 'VicFleet' to assume responsibility for those agencies that were previously contracted to LeasePlan. The contract of GE Fleet Services had been successfully renewed prior to expiry. A DTF high level review in September 2001 showed that the government seemed to pay more and receive less service when compared with best practice fleet management services.

Advantages of public-private partnership (sale and lease-back)

The main advantages of PPP in this project are (1) savings from sale of vehicles: the capital tied up in the vehicle fleet can be released for better use elsewhere; (2) cost avoidance: sale and lease-back reduces capital spending on acquiring new/replacement vehicles and eliminates recurrent expenditure in management and maintenance of the vehicles in use; and (3) innovation: government can benefit from advanced technology and management practices introduced by the private sector.

Challenges under the sale and lease-back arrangement

The challenges/issues stemming from the sale and leaseback arrangement include:

1 The liabilities of government on early termination: the government faced excessive termination payments for the first three years of its term;

2 The absorption of vehicle residual value risk by government: the state was required to fund the deficit between the actual value of vehicles sold and the previously consented residual value of the vehicles through increased rentals on new leases. The government suffered potential losses of around AU\$88.3 million up to June 2001 because of the downward movements in used car prices;

3 Transfer of minimal risk to CBA: the lease payments were designed to ensure the CBA's achievement of a required return rate and the financing risk remained with the government;

4 Lack of a business case: during a period for LeasePlan assuming responsibility to provide fleet management services, the DTF had not developed a business case to support the activity. In particular, the DTF had not evaluated with cost benefit analysis whether the services could be effectively managed by another private provider or by a government agency.

Lessons learned

When considering enhancing the role of the private sector to provide vehicle fleet management services, sale and lease-back arrangements could be taken into account. The following points should be considered.

1 Multiple contracts: the government vehicle fleet could be divided into different groupings (i.e. geographically or functionally) to ensure that a competitive environment was maintained. In addition, it is better to ascertain that competitors, as a contingency, could step in to take over a failing service provider.

2 Risk sharing and management: the private sector should be contracted to share and manage the risks related to financing, fleet operation and migration from internal fleet operation to external fleet management. It is better for the government to consider which risks could be transferred to the private partner for more effective management.

3 Specification of service requirements and service levels: it should define clearly the services to be provided in agreements.

4 Performance monitoring: contract administrators should assess the service delivery performance of contractors against the outcome specifications in the contract.

5 Staffing: the Victoria experience had little impact on staff due to the preponderance of user-drivers, therefore, it is essential to consider carefully the staffing implications.

6 Renewal of fleet management contract: government should allow enough time to assess the best service delivery options prior to considering renewal of a service contract.

7 Contract arrangements for dispute resolution: agreements should specify in detail the form of default notices, the rights and obligations of the respective parties, the dispute resolution process, and the terms of contract termination.

8 Other private sector involvement choices: it is not a definite conclusion that the arrangements of PPP approach would be the best option. Other alternatives, such as outsourcing or privatization, might provide better solutions under different situations.

Comparisons of the case studies

Similarities

Most of the selected case studies in PPP in the UK and Australia indicated that appropriate risk allocation and risk sharing is a key to PPP success. In general terms, this means allocating each risk to the party best able to manage it and this helps to reduce individual risk premiums and the overall cost of the project because the party should be able to manage a particular risk under such the best situation. In addition, many selected case studies in PPP revealed that taking an appropriate economic appraisal, i.e. thorough and realistic cost and benefit assessment, of different internal and external service options could help to achieve the most cost-effective outcome and a suitable assessment of private sector tenders.

Differences and highlights

Based on the aforesaid case studies in PPP, it can be concluded that many PPP projects were implemented successfully in the UK and Australia. Their lessons learned are summarized and highlighted in Table 13.2.

Table 13.2 Lessons learned from the six selected case studies in PPP in the UK and Australia

Lessons learned	Case Study 1	Case Study 2	Case Study 3	Case Study 4	Case Study 5	Case Study 6
1. Specification of service requirements and service levels				√	√	√
2. Value for money testing	√	√				
3. Achieving affordability	√	√				
4. Consultation	√	√				
5. Evaluation of service delivery options				√	√	
6. Performance review and monitoring				√		√
7. Staffing quality assurance				√		√
8. Contract arrangements for dispute resolution				√		√
9. Inclusion of property		√				
10. Well articulated estate rationalization plan			√			
11. Requirement for investment in new facilities			√			
12. Focus on a core facility			√			
13. Combination of separate court buildings and integration of non-core services				√		
14. Change management				√		
15. Continuous technology improvement				√		
16. Supply of source code and intellectual property for IT services				√		
17. Well-established practices					√	
18. Multiple contracts						√
19. Risk sharing and management						√
20. Renewal of fleet management contract						√
21. Other private sector involvement choices						√

Chapter 14

Case studies in joint venture

This chapter shows how joint venture (JV) is implemented in three different nations. A total of three case studies in JV in the USA, the UK and Hong Kong are selected for demonstration purposes.

Case study 1

Tsing Ma Bridge, Hong Kong (extracted from the public domain website of http://en.wikipedia.org/wiki/Tsing_Ma_Bridge and http://www.cse.polyu. edu.hk/~ctbridge/case/tsingma.htm).

Background

The Tsing Ma Bridge is a suspension bridge in Hong Kong, China and is named after two of the islands it connects, *Tsing* Yi and *Ma* Wan. The Tsing Ma Bridge is the seventh longest-span suspension bridge in the world. The client was the Highways Department of the Hong Kong SAR Government; the main contractor was Anglo Japanese Construction JV; and the engineer and designer was Mott MacDonald Hong Kong Ltd. It has two decks and carries both road and rail traffic. The main span of the bridge is 1,377 metres (4,518 ft) long and 206 metres (676 ft) high. The span is the largest of all bridges in the world that carry rail traffic. The 41 metres (135 ft) wide bridge deck carries six lanes of automobile traffic, three lanes in each direction. There are two rail tracks in the lower level. In addition, there are two sheltered carriageways on the lower deck for both maintenance access and as backup for traffic in case typhoons strike Hong Kong. Although road traffic would need to be closed in that case, trains could still get through in either direction. Table 14.1 shows the details of this JV project.

Construction and operations

There is a fully suspended main span in the bridge. It is supported by two portal-braced and reinforced-concrete towers. The bridge deck is suspended

Table 14.1 Details of the Tsing Ma Bridge Project (reproduced from the website (public domain) of http://en.wikipedia.org/wiki/Tsing_Ma_Bridge)

Official name	Tsing Ma Bridge
Client	Highways Department of the Hong Kong government
Engineer and designer	Mott MacDonald Hong Kong Limited
Main contractor	Anglo Japanese Construction JV
Construction date	May 1992
Completion date	May 1997
Opening date	27 April 1997
Construction cost	HK$7.2 billion
Toll	HK$30 (cars)
Carries	6 lanes of roadway (upper); 2 MTR rail tracks; 2 lanes of roadway (lower)
Crosses	Ma Wan Channel
Locale	Ma Wan Island and Tsing Yi Island
Design	Double-decked suspension bridge
Longest span	1377 metres (4518 ft)
Width	41 metres (135 ft)
Height of towers	206 metres (675 ft)
Vertical clearance	62 metres (203 ft)

from two main cables, which pass over the main towers and are secured into massive concrete anchorages at each end on Tsing Yi and Ma Wan islands. The main cables are retained in steel saddles at the top of each tower to supply the cable geometry and to transfer their loads to the towers.

The two towers are composed of twin columns. When the height increases, they will taper with the legs constructed by continuous slipforming. The two main cables were made up of some 30,000 tonnes of specialized, high-tensile, galvanized steel wire. They were built up wire-by-wire by the in situ aerial spinning process, with each cable spun and wrapped from 32,000 wires. The bridge deck was formed by 96 fabricated steel sections measuring some 40m wide by 7.3m deep. It is supported at intervals of 18m throughout the length by suspenders connected to the main cables.

The construction of the bridge was carried out by a Costain/Mitsui/Trafalgar House joint venture. The bridge was constructed starting from May 1992 and ending at May 1997, with a cost of HK$7.2 billion. The Lantau Link was opened on April 27, 1997. The Tsing Ma Bridge links Tsing Yi Island on the east to Ma Wan island on the west over Ma Wan Channel and is part of the Lantau Link. With two long-span bridges linking

Figure 14.1 Tsing Ma Bridge

the New Territories and Lantau Island, it eventually leads to the Hong Kong International Airport on Chek Lap Kok via North Lantau Highway. The other bridge is the Kap Shui Mun Bridge; it links Ma Wan to Lantau Island over Kap Shui Mun. The two bridges are connected over Ma Wan by Ma Wan Viaduct.

The Tsing Ma Bridge has been a vital gateway to Lantau Island. It is Route 8 expressway, which connects the Lantau Link, the West Kowloon expressway, Cheung Sha Wan and Shatin. The rail line is part of MTRCL's Tung Chung Line and Airport Express. The bridge, together with other highway, bridge and tunnel connections in the area, is part of the Tsing Ma Control Area under the Tsing Ma Control Area Ordinance (Cap. 498) in Hong Kong Law. The control area has been managed by Tsing Ma Management Ltd since commencement. The traffic management system of control area was developed by Delcan Corporation of Toronto. Special regulations and by-laws are carried out in the area. The bridge is also closely monitored by the Wind and Structural Health Monitoring System (WASHMS). Surveillance cameras are installed over the bridge to record traffic conditions. The maximum speed limit on the bridge is always 80 km/h (50 mph) for automobiles. Such speed limits may be lowered when there are road works or strong winds. Traffic may also be directed to the sheltered carriageways on the lower deck when there are very strong winds. There is no pedestrian pavement on the bridge, and parking is also prohibited on the bridge.

Design

The bridge was designed by Mott MacDonald. The objectives of the wind tunnel studies were to demonstrate the safety of the structure under construction and once completed, both with respect to aerodynamic stability and the possible effects of extreme typhoon wind speeds. Another objective was to provide dynamic response data at some key locations to compare with full scale data from the continuous monitoring programme conducted by the Highways Department of Hong Kong.

A 1-to-80 scale section model of the deck in the erection stage, and a 1-to-400 scale full aeroelastic model of the whole bridge were constructed. It is a Monte-Carlo simulation of the typhoon wind climate. The full model was tested in various stages of construction in turbulent boundary layer flow. It was completed with the local topography so as to model the wind conditions at the site. The model tests identified critical stages of erection that permitted the construction schedule of the bridge to be tailored to prevent interference from the typhoon season. The comparison of model test results and the full scale monitoring can help engineers to better understand the behaviour of long-span bridges in wind and to improve current design methods.

Major components

1 Bridge tower foundations: one tower situated at Wok Tai Wan of Tsing Yi side and the other on a man-made island 120m from the coast of Ma Wan Island. Both towers are 206m above sea level and they are founded on relatively shallow bedrock. The towers are two-legged with trusses at intervals. They are in the form of portal beam design. The legs were constructed with high-strength concrete of 50 MPa (concrete grade 50/20) strength with a usage of a slipform system in a continuous operation.

2 Anchorages: the pulling forces in the main suspension cables is taken up by large gravity anchorages located at both ends of the bridge. They are massive concrete structures deeply seated on bedrock on the landside of Tsing Yi and Ma Wan island. The total weight of concrete used in the Tsing Yi anchorage and Ma Wan Anchorage is 200,000 tonnes and 250,000 tonnes respectively.

3 Main cables: the cables were constructed by an aerial spinning process. The process involved drawing wires from a constant-tension supply, and pulling loops of these wires from one anchorage to the other, passing a 500-tonne cast-iron saddle on top of each bridge tower seating the cable. There were a total of 70,000 galvanized wires of

5.38 mm diameter placed and adjusted to form the two 1.1m diameter main cables.

4 Suspended deck: the steelwork for the deck structure was fabricated in the UK and Japan. After delivery, they were further processed and assembled in Dongguan of China into standard deck modules. There were 96 modules in total and each of them is 18m long and weighs about 480 tonnes. These deck modules were brought to the site by specially designed barges and raised into the deck position by a pair of strand jack gantries that could move along the main cable.

5 Approach span on Tsing Yi side: similar in form and cross-section to the suspended deck, but the approach span was supported on piers rather than cables. The first span was assembled on the ground and raised into position using strand jacks. Further erection was then cantilevered in smaller sections by using derrick cranes stationed on the deck level. An expansion joint which allowed for a maximum thermal movement of ±835mm was also provided and located inside the approach span section.

Lessons learned

The contract of CJV is quite flexible in that it permitted the company to negotiate with its prospective partners without any need to argue about valuation methods. The CJV contract also allowed E-Smart to obtain the management rights that it desired. By doing so, the tax advisors of the company felt more comfortable with the CJV for overall international tax planning.. In addition, it helps to spread and reduce project risk. It is concluded that JV, when implemented appropriately, can generate a good practical model for different companies to work together more cooperatively and effectively, thus enhancing the pooling of knowledge, technology and resources.

Case study 2

A 'smart card' wireless security technology cooperative joint venture (CJV), USA (extracted from the website of *China Business Review*, the magazine of the US-China Business Council http://www.chinabusinessreview.com/public/0501/folta.html) (permission has been obtained for both print and online use from the US-China Business Council).

Background

A US company, e-Smart Technologies, Inc., mainly operates wireless security technology for smart cards. It signed a CJV agreement with two Chinese companies in early 2004. Under the agreement, the CJV would operate countrywide value-added networks constituting e-Smart's operating platform and its multi-application, secure ID and payment smart cards. In addition, the CJV can market the system and technologies to the government and financial sectors, together with maximizing the technology's usage in as many fields as possible.

One of the Chinese CJV partners is majority owned and managed by an entity of the Ministry of Information Industry (MII). The other is composed of Chinese media and public relations personnel. E-Smart owns half of this joint venture (the maximum allowed by law for a value-added service venture), and the two Chinese companies own 30% and 20% respectively. After issuing all required permits and licences, e-Smart contributes roughly $3 million in capital to the CJV over time. This $3 million represents 100% of the CJV's registered capital, but the venture may be expanded. E-Smart owns the exclusive licenses to provide and operate the system and technologies in China. And it can receive 20% of the CJV's gross operating income. The Chinese parties can use their relationships with the authorities to obtain the needed licenses and approvals. In addition, they can participate in market promotion and negotiations with customers, address network infrastructure issues and help to obtain financing. The board and operating management team are determined by the parties in consultation with each other.

Lessons learned

According to e-Smart executives, there was nothing to do for specific telecom sector with the CJV structure. The executives viewed that the CJV gave different companies the flexibility required to handle the regularly changing circumstances, regulations and laws that one must compete with when doing business in China. The flexibility of a CJV contract allowed the company to negotiate with its prospective partners with no need to argue about valuation methods. The CJV form also permitted e-Smart to obtain the management rights that it desired. As a result, the tax advisors of the company became more comfortable with the CJV for overall international tax planning.

The CJV required a detailed agreement and e-Smart believed that this detail helped to protect its technology if it is coupled with all of the ancillary agreements and a clear licensing agreement. A company executive stated that all of the agreements made it very clear that the technology was not being transferred and that the ownership remained in the hands of e-Smart alone. While the CJV is usually more time-consuming and complicated to negotiate in the beginning, it is this complexity that becomes a benefit. The companies involved are forced to think through all of the possible problems that may occur in the future and deal with them up front. The result is a smoother relationship with partners and a good plan for the operation of the CJV.

Case study 3

Broadgreen Hospital redevelopment, Liverpool, UK: An integrated health project (extracted from the website of http://www.nhs-procure21.gov.uk/news/downloads/93/broadgreenhospitalredevelopmentliverpoolcasestudyrevb_000.pdf) (permission has been obtained for both print and online use from the Department of Health, UK).

Background

The Broadgreen Hospital redevelopment project is the largest and highest value project delivered through the Department of Health's ProCure21 framework. The construction cost was £63m. This is one of the integrated health projects (IHP) and it is a JV between Sir Robert McAlpine Ltd and Norwest Holst Construction Ltd. The scheme was constructed in seven phases, and it was constructed within budget and on time. It provides extensive new-build accommodation and major changes and extensions to the cardiothoracic centre's operating department, together with the alteration and reuse of surplus accommodation. The completed development provides the Royal Liverpool and Broadgreen University Hospitals NHS Trust with a devoted treatment centre, including orthopaedic, urological and general surgical facilities, wards and a day surgery unit. The cardiothoracic centre was given an enlarged operating department, catheter laboratories, a new critical care facility, out-patient department, acute wards and health records department.

The successful functioning of the development and its affordability was primarily based on the careful disposition and design of highly serviced accommodation shared by the two trusts (a new imaging department, pharmacy, pathology department and a new main entrance). The scheme exploits its sloping site and makes extensive use of natural light, courtyard landscaping and colour to upgrade the patient environment, aid way-finding and create a high-quality working environment for staff. It also includes some new features including an orthopaedic 'barn' operating theatre which caters for four operating tables within an open-plan area.

Effective design development leading through to guaranteed maximum price

After the initial meeting with the trusts was held, a detailed programme was prepared. It clearly showed the processes to be followed to achieve a GMP together with team members concerned and timescales to be realized. This programme was backed up with details of the personnel who would be involved including an assessment of the number of days for their involvement and their hourly rate, costs for surveys and site investigations.

Based on this information, the trust was asked to estimate the costs required in the period up to GMP. The design effectively began with the healthcare planner, Directions Consultancy, which worked with the trusts to set up a definitive brief and schedule of accommodation. Directions Consultancy took the approach that service delivery was vital. It therefore challenged preconceived ideas by the user groups and, in some cases, proposed operational policies.

This approach had the effect of adding area in some departments. It succeeded in reducing the overall area by 10% from the initial accommodation schedule. Directions Consultancy worked closely with architects, Nightingale Associates, during the design process. As the design was developed, other supply chain members were introduced at the point when they could add most value. Then, it introduced mechanical and electrical (M&E) installer, EMCOR, and their extended specialist supply chain, to ensure that the M&E services were fully integrated into the design. To match with design, the cost model was prepared and the initial risk workshop and subsequent risk management schedule was implemented.

The cost modelling allowed value management techniques to be put in place with budgets being given for the design elements. This permitted the designers to progress with full knowledge of their cost constraints. Throughout this whole period up to GMP, a major challenge was to settle the agenda of the two trusts. A key element in keeping a large number of interested parties fully informed of the design development was the use of the 4Projects collaboration tool. After some initial problems, it proved to be an excellent medium to share all information. Although it was complicated for the period up to GMP, the client was successful in remaining within the cost estimate given at the beginning for the period up to GMP.

Effective management of construction and commissioning

The project was planned in detail and an in-depth phasing strategy was prepared and agreed with the trusts. Trust input identified the areas were to be prioritized and enabled them to plan a strategy for the refurbishment areas in good time. Detailed planning was the key to the success of the project and the involvement of the integrated and dedicated team in the process could make sure that the trusts were able to carry out their own planning for equipping and commissioning the completed areas. It also allowed early warning of problems so that it could avoid delay to the overall completion.

The contract programme was produced by using information from all concerned parties and, in particular, incorporated the periods identified by the major and specialist work package contractors. It was maintained as a living document, being regularly reviewed to ensure that the logic remained relevant, to consider any changes to the work and to give advice to the trusts on the programme implications of changes under consideration. This approach to planning and programming guaranteed that all the departmental opening dates were met and allowed the additional works instructed by the trusts to be incorporated within the project period. On all the sites, health and safety matters were paramount and the JV partners of IHP have an excellent track record in this respect.

Problems overcome

The design solution not only meets the various needs of the two clients but it also successfully links the two trusts' existing accommodation, which is located at opposing corners of the hospital grounds and across a sloping site over four levels. The precise disposition and internal planning of shared facilities were critical to the scheme's success. They were set up after a comprehensive appreciation of user needs was gained by developing and testing a series of alternative solutions. Norwest Holst recognized from the outset that hospitals and construction seldom mixed, and planned to provide an 'invisible' presence on site. This was achieved by establishing a separate site entrance, segregating the hospital completely from construction activities. Specific problems were overcome by providing additional methods for sound-proofing and early warning procedures to stop works if essential.

Innovation

Since the barn-like operating theatre was a relatively new concept and there was little experience available in the UK, it was a particular challenge to the services team. No guidelines were available at that time. Subsequently, one of the leading experts in the country, Malcolm Thomas, used Broadgreen as the basis for guidelines to be published in the future. A specialist subcontractor, Howarth Airtech, was appointed to underwrite the performance of the airflow systems that were designed to avoid any air mixing across the four operating tables, and thereby isolating one area from the next.

Having an open consultation process between all parties, it was possible to introduce some innovations into the project that would benefit all parties. Two particularly beneficial new methods for the mechanical and electrical services were the use of pre-wired distribution boards and the use of flexi-shield cable in all areas. The former can improve quality and saved installation time, and the latter can considerably reduce the requirement for electrical containment. It is difficult to measure value for money and it should be viewed widely that it should not be measured by monetary terms only. The Broadgreen project is a high-quality development with world-class facilities at an affordable cost.

Lessons learned

The team has held two project reviews, one for the construction process and the other for the design process. Both forums proved positive and were an honest exchange of views, which can benefit future projects. 4Projects, the collaboration tool-kit used on the project, was a learning curve for many and would become more extensive and will be introduced to a project from the outset. It aims to persuade the more unwilling parties of the benefits and ensure that the right level of training and support is given. To minimize the possibility of misinterpretation of drawings by non-technical users, it would present three-dimensional layouts for future projects.

The Broadgreen Hospital redevelopment project has good design, high quality, innovation, delivery certainty, fully integrated team and an excellent safety record. This has resulted in the provision of world-class facilities.

Comparisons of the case studies

Differences and highlights

Table 14.2 summarizes the lessons learned from the three selected case studies in JV. Nine lessons learned have been identified and it should be noted that they are completely different. Like PPP and other construction projects, the distinguishing lessons learned from these case studies reflect that JV projects are unique, and not generic, in nature so that the combined efforts of different companies, especially for those companies adopting international joint venture to procure their construction projects, are probable to bring different project outcomes and diversified lessons learned.

Table 14.2 Lessons learned from the three selected case studies in JV

Lessons learned	Case Study 1	Case Study 2	Case Study 3
1. Knowledge and technology transfer	√		
2. Spread and reduction of risk	√		
3. Flexibility required to handle the regularly changing circumstances, regulations and laws		√	
4. Smoother relationship with partner		√	
5. Training and support			√
6. Effective design development leading trough to GMP			√
7. Effective management of construction and commissioning			√
8. Innovations			√
9. Fully integrated teams			√

Appendices

Goals and objectives

The RC Charter needs an understanding of individual specific objective of all stakeholders involved in this project.

Company: _____

Functional group– ❏ Client ❏ Consultant
                                ❏ Main Contractor ❏ Subcontractor

Please list the goals and objectives of you or your organization in this project:

Details:

1. _____
2. _____
3. _____
4. _____
5. _____
6. _____
7. _____
8. _____
9. _____
10. _____
11. _____
12. _____
13. _____
14. _____
15. _____
16. _____
17. _____
18. _____
19. _____
20. _____
21. _____
22. _____
23. _____
24. _____

Appendix 2 Pro forma worksheet 2: Creating Ideas for Successful RC

Creating ideas for a successful RC

Company: _____

Sub-group number (for cross-disciplinary teams):_____

Sub-group topic:_____

Please brainstorm any possible ideas/suggestions to make RC work in this project.

1. _____
2. _____
3. _____
4. _____
5. _____
6. _____
7. _____
8. _____
9. _____
10. _____
11. _____
12. _____
13. _____
14. _____
15. _____
16. _____
17. _____
18. _____
19. _____
20. _____
21. _____
22. _____
23. _____
24. _____

Appendix 3 Pro forma worksheet 3: Potential Hindrances of RC

Potential hindrances of RC

Company: _____

Functional group–   ❏ Client   ❏ Consultant
                     ❏ Main Contractor ❏ Subcontractor

Please identify potential hindrances/obstacles which may emerge from you or your colleagues that could impede successful completion of the project.

1. _____
2. _____
3. _____
4. _____
5. _____
6. _____
7. _____
8. _____
9. _____
10. _____
11. _____
12. _____
13. _____
14. _____
15. _____
16. _____
17. _____
18. _____
19. _____
20. _____
21. _____
22. _____
23. _____
24. _____

Appendix 4 Pro forma worksheet 4: Overcoming Hindrances of RC

Overcoming hindrances of RC

Company: _____

Functional group–   ❑ Client   ❑ Consultant
                    ❑ Main Contractor ❑ Subcontractor

Please examine the identified potential hindrances and then suggest recommendations to overcome them.

1. _____
2. _____
3. _____
4. _____
5. _____
6. _____
7. _____
8. _____
9. _____
10. _____
11. _____
12. _____
13. _____
14. _____
15. _____
16. _____
17. _____
18. _____
19. _____
20. _____
21. _____
22. _____
23. _____
24. _____

Appendix 5 Pro forma worksheet 5: Developing RC Strategies for Implementation

Developing RC strategies for implementation

Company: _____

Sub-group number (for cross-disciplinary teams):_____

Sub-group topic:_____

Please develop any possible RC strategies for implementation in this project.

1. _____
2. _____
3. _____
4. _____
5. _____
6. _____
7. _____
8. _____
9. _____
10. _____
11. _____
12. _____
13. _____
14. _____
15. _____
16. _____
17. _____
18. _____
19. _____
20. _____
21. _____
22. _____
23. _____
24. _____

Appendix 6 Pro forma worksheet 6: Issue Escalation

Company: _____

Issue escalation

Sub-group number: _____

The timely resolution of issues is vital to the effective management of a project. This issue resolution tree allows parties to know the people who are empowered to make decisions at each operational level. Please enter a name or names against each of the following boxes.

Company Name	Client	Architectural Consultant	Structural Consultant	Building Services Consultant	Quantity Surveyor	Main Contractor	Electrical & Fire Services	MVAC Installation	ELS Works and Excavator
1 (Senior Management)									
2									
3									
4 (Front Line)									

Notes:

Code 4	Code 3	Code 2	Code 1
Site Foreman	Site Agent/Clerk of Works	Head Office Project Manager	Director

Appendix 7 Partnering Performance Monitoring Matrix

RC Performance Monitoring Matrix

Company name: _____

Staff name: _____

Position: _____

Scoring system:

5	4	3	2	1
Very satisfactory	Satisfactory	Average	Unsatisfactory	Very unsatisfactory

1st shared objective	Score 評級	Apr 2008	May 2008	Jun 2008	Jul 2008	Aug 2008	Sep 2008	Oct 2008	Nov 2008	Dec 2008	Jan 2009	Feb 2009	Mar 2009
	Sample size												
	5												
	4												
	3												
	2												
	1												

2nd shared objective	Score 評級	Apr 2008	May 2008	Jun 2008	Jul 2008	Aug 2008	Sep 2008	Oct 2008	Nov 2008	Dec 2008	Jan 2009	Feb 2009	Mar 2009
	Sample Size												
	5												
	4												
	3												
	2												
	1												

3rd shared objective	Score 評級	Apr 2008	May 2008	Jun 2008	Jul 2008	Aug 2008	Sep 2008	Oct 2008	Nov 2008	Dec 2008	Jan 2009	Feb 2009	Mar 2009
	Sample Size												
	5												
	4												
	3												
	2												
	1												

Appendix 8 Pro forma worksheet 7: Perceived Benefits of RC

Perceived benefits of RC

Please suggest the perceived benefits that you had experienced in this RC project.

Company: _____

Sub-group number (for cross-disciplinary teams): _____

1. _____
2. _____
3. _____
4. _____
5. _____
6. _____
7. _____
8. _____
9. _____
10. _____
11. _____
12. _____
13. _____
14. _____
15. _____
16. _____
17. _____
18. _____
19. _____
20. _____
21. _____
22. _____
23. _____
24. _____

Appendix 9 Pro forma worksheet 8: Benefits of RC Identified from Academic Literature

Company name:_____

Name of participant: _____

Sub-group number: _____

Position: _____

Please rate the following benefits that you have experienced in this RC project.	Not significant	Slightly significant	Moderately significant	Very significant	Extremely significant
1. Improved relationship amongst project participants	❏	❏	❏	❏	❏
2. Improved communication amongst project participants	❏	❏	❏	❏	❏
3. More responsive to the short-term emergency, changing project or business needs	❏	❏	❏	❏	❏
4. Reduction in dispute	❏	❏	❏	❏	❏
5. Better productivity was achieved	❏	❏	❏	❏	❏
6. A win-win attitude was established amongst the project participants	❏	❏	❏	❏	❏
7. A long-term trust relationship was achieved	❏	❏	❏	❏	❏
8. Improved corporate culture amongst project participants	❏	❏	❏	❏	❏
9. Reduction in litigation	❏	❏	❏	❏	❏
10. Improved conflict resolution strategies	❏	❏	❏	❏	❏
11. Reduction in monetary claims	❏	❏	❏	❏	❏
12. Reduction in variations	❏	❏	❏	❏	❏
13. Cost saving was achieved	❏	❏	❏	❏	❏
14. Faster construction time was achieved	❏	❏	❏	❏	❏
15. Better workmanship, less rework	❏	❏	❏	❏	❏
16. Increased opportunity for innovation	❏	❏	❏	❏	❏
17. Project risks were shared more equitably amongst project participants	❏	❏	❏	❏	❏
18. Other:_____	❏	❏	❏	❏	❏
19. Other: _____	❏	❏	❏	❏	❏
20. Other: _____	❏	❏	❏	❏	❏

Appendix 10 Pro forma worksheet 9: Perceived Major Difficulties in Implementing RC

Perceived major difficulties in implementing RC

Please list the significant difficulties that you had experienced in this RC project.

Company: _____

Sub-group number (for cross-disciplinary teams): _____

1. _____
2. _____
3. _____
4. _____
5. _____
6. _____
7. _____
8. _____
9. _____
10. _____
11. _____
12. _____
13. _____
14. _____
15. _____
16. _____
17. _____
18. _____
19. _____
20. _____
21. _____
22. _____
23. _____
24. _____

Appendix 11 Pro forma worksheet 10: Major Difficulties in Implementing RC Identified from Academic Literature

Company name: _____

Name of participant: _____ Sub-group number: _____

Position: _____

Please rate the following difficulties that you have experienced in this RC project.	Not significant	Slightly significant	Moderately significant	Very significant	Extremely significant
1. Parties were faced with commercial pressure which compromised the RC attitude	❏	❏	❏	❏	❏
2. The parties had little experience with the RC approach	❏	❏	❏	❏	❏
3. Uneven levels of commitment were found amongst project participants	❏	❏	❏	❏	❏
4. Risks or rewards were not shared fairly	❏	❏	❏	❏	❏
5. The concept of RC was not fully understood by the participants	❏	❏	❏	❏	❏
6. Dealing with large bureaucratic organizations impeded the effectiveness of RC	❏	❏	❏	❏	❏
7. Conflicts arose from misalignment of personal goals with the project goals	❏	❏	❏	❏	❏
8. Participants were conditioned in a win-lose environment	❏	❏	❏	❏	❏
9. Parties did not have proper training on RC approach	❏	❏	❏	❏	❏
10. The RC relationship created a strong dependency on other partners	❏	❏	❏	❏	❏
11. Some participants did not accept RC as a long-term way of doing business	❏	❏	❏	❏	❏
12. Continuity of open and honest communication was not achieved	❏	❏	❏	❏	❏
13. Parties failed to build a true relationship of trust	❏	❏	❏	❏	❏
14. Key subcontractors had not been included in the process	❏	❏	❏	❏	❏
15. The owner and slave concept was alive in the project	❏	❏	❏	❏	❏
16. The parties failed to share information	❏	❏	❏	❏	❏
17. Top management just gave lip-service to the RC approach	❏	❏	❏	❏	❏
18. Issues and problems were allowed to slide and escalate	❏	❏	❏	❏	❏
19. Other: _____	❏	❏	❏	❏	❏
20. Other: _____	❏	❏	❏	❏	❏
21. Other: _____	❏	❏	❏	❏	❏

Appendix 12 Pro forma worksheet 11: Perceived Critical Success Factors for Adopting RC

Perceived critical success factors for adopting RC

Please list the perceived critical success factors for adopting RC that you had experienced in this RC project.

Company: _____

Sub-group number (for cross-disciplinary teams): _____

1. _____
2. _____
3. _____
4. _____
5. _____
6. _____
7. _____
8. _____
9. _____
10. _____
11. _____
12. _____
13. _____
14. _____
15. _____
16. _____
17. _____
18. _____
19. _____
20. _____
21. _____
22. _____
23. _____
24. _____

Appendix 13 Pro forma worksheet 12: Critical Success Factors for Adopting RC Identified from Academic Literature

Company name: _____

Name of participant: _____ Sub-group number: _____

Position: _____

Please rate the following factors that are critical in affecting the success of this RC project.	Not significant	Slightly significant	Moderately significant	Very significant	Extremely significant
1. Establishment and communication of conflict resolution strategy	❑	❑	❑	❑	❑
2. Commitment to win-win attitude	❑	❑	❑	❑	❑
3. Regular monitoring of RC process	❑	❑	❑	❑	❑
4. Clear definition of responsibilities	❑	❑	❑	❑	❑
5. Mutual trust among team members	❑	❑	❑	❑	❑
6. Willingness to eliminate non-value-added activities	❑	❑	❑	❑	❑
7. Early implementation of RC process	❑	❑	❑	❑	❑
8. Willingness to share resources among project participants	❑	❑	❑	❑	❑
9. Ability to generate innovative ideas	❑	❑	❑	❑	❑
10. Subcontractors' involvement during the RC process	❑	❑	❑	❑	❑
11. RC principles fully supported and committed at all levels of management for each stakeholder organization	❑	❑	❑	❑	❑
12. Fair deal from the partners	❑	❑	❑	❑	❑
13. Establishment of mutual goals among project participants	❑	❑	❑	❑	❑
14. Open airing of problems and opinions during RC process	❑	❑	❑	❑	❑
15. Appointment of a RC team leader or champion to ensure that the RC principles did not slip out of focus	❑	❑	❑	❑	❑
16. Other: _____	❑	❑	❑	❑	❑
17. Other: _____	❑	❑	❑	❑	❑

Appendix 14 Pro forma worksheet 13: Overcoming Hindrances

Overcoming hindrances

Please examine the identified 'potential hindrances' and then suggest recommendations to overcome them. ('How to solve...' questions)

Company: _____

Sub-group number (for cross-disciplinary teams): _____

1. _____
2. _____
3. _____
4. _____
5. _____
6. _____
7. _____
8. _____
9. _____
10. _____
11. _____
12. _____
13. _____
14. _____
15. _____
16. _____
17. _____
18. _____
19. _____
20. _____
21. _____
22. _____
23. _____
24. _____

Notes

1 Different researchers use the terms 'Relational Contracting', 'Relational Contracts', 'Relationship Contracting', and 'Relationship Contracts' with the same meaning and therefore they are considered interchangeable in this book. Since it seems that 'Relational Contracting' is more frequently used in the latest publications, it is used throughout the whole book to maintain consistency.

2 Content analysis is often used to determine the main facets of a set of data, by simply counting the number of times an activity occurs or a topic is mentioned (Fellows and Liu, 2008). The initial step in content analysis is to identify the materials to be analysed. The next step is to determine the form of content analysis to be employed, including qualitative, quantitative or structural; the choice is dependent on the nature of the research project. The choice of categories will also depend upon the issues to be addressed in the research if they are known. In qualitative content analysis, emphasis is on determining the meaning of the data (grouping data into categories), while quantitative content analysis extends the approach of the qualitative form to yield numerical values of the categorized data (frequencies, ratings, ranking, etc) which may be subjected to statistical analyses. Comparisons can be made and hierarchies of categories can be examined.

3 It seems that there is no research study on performance measures for PPP projects.

4 Unlike structured approach, which is typically composed of one initial partnering workshop, one or more interim partnering workshops, and one final partnering wrap-up workshop, unstructured approach is one implementing partnering spirit only, such as emphasizing mutual trust, win-win philosophy, fair deal and top management commitment, but without launching any partnering workshops.

5 A strategic alliance is a formal relationship between two or more parties to pursue a set of agreed goals or to meet a critical business need whilst remaining independent organizations (Wikipedia, 2008). Parties may provide the strategic alliance with resources such as products, distribution channels, manufacturing capability, project funding, capital equipment, knowledge, expertise or intellectual property. The alliance is cooperation or collaboration which aims for a synergy where each partner hopes that the benefits from the alliance will be greater than those from individual efforts. The alliance often involves technology transfer, economic specialization, shared expenses and shared risk.

6 Co-marketing alliances can be defined as a concern of the achievement of market growth and new market entry and expansion (Harrigan, 1985; Kelly, 1990).

References

Abrahams, A., and Cullen, C. (1998). 'Project alliances in the construction industry.' Australian Construction Law Newsletter. Oct/Nov. 31–36.

Achrol, R., Scheer, L., and Stern, L.W. (1990). 'Designing successful trans-organizational marketing alliances.' Marketing Science Institute, Cambridge, MA.

Adler, L. (1966). 'Symbiotic marketing.' Harvard Business Review, 44, November-December, 59–71.

Abudayyeh, O. (1994). 'Partnering: A team building approach to quality construction management.' Journal of Management in Engineering, ASCE, 10(6), 26–29.

Adnan, H., and Morledge, R. (2003). 'Application of Delphi method on critical success factors in joint venture projects in the Malaysian construction industry.' Proceedings of CITC-II Conference, Hong Kong, December 10–12, 2003.

Akintoye, A., Beck, M., Hardcastle, C., Chinyio, E., and Asenova, D. (2001). 'The financial structure of private finance initiative projects.' Proceedings of the 17th ARCOM Annual Conference, Salford University, Manchester, 1, 361–369.

Albanese, R. (1994). 'Team-building process: key to better project results.' Journal of Management in Engineering, ASCE, 10(6), 36–44.

Alchimie Pty Ltd and Phillips Fox Lawyers. (2003). 'Project alliances: An overview.'

Al-Meshekeh, H.S., and Langford, D.A. (1999). 'Conflict management and construction project effectiveness: A review of the literature and development of a theoretical framework.' Journal of Construction Procurement, 5(1), 58–75.

Alsagoff, S.A., and McDermott, P. (1994). In Rowlinson, S. (ed.) 'Relational contracting: A prognosis for the UK construction industry?' Proceedings of CIB W92, Procurement Systems – East Meets West, University of Hong Kong, 11–19.

American Arbitration Association. (1996). 'A Guide to Partnering in the Construction Industry – Building Success for the 21st Century.' Report of the Dispute Avoidance and Resolution Task Force of the American Arbitration Association.

APM-HK. (2003). 'Partnering Guidelines for Construction Projects in Hong Kong.' Association for Project Management Hong Kong, Partnering Specific Interest Group, June, 48 pages.

Asian Development Bank. (2007). 'Application of public-private partnerships in urban rail-based transportation project.' Mass Transit Railway Corporation Limited, Hong Kong.

Associated General Contractors of America. (1991). 'Partnering: a concept for success.' Associated General Contractors of America, Arlington, VA.

Australian Constructors Association. (1999). 'Relationship contracting – Optimising project outcomes.' Australian Constructors Association, Sydney.

Back, W.E., and Sanders S.R. (1996). 'Partnering in a unit price environment.' Project Management Journal, 27(2), 18–25.

Bates, G.D. (1994). 'Partnering is small packages.' Journal of Management in Engineering, ASCE, 10(6), 22–23.

Bayliss, R.F. (2000). 'Project partnering – a case study on MTR Corporation Ltd's Tseung Kwan O Extension.' Proceedings of the Millennium Conference on Construction Project Management – Recent Developments and the Way Forward 2000, Hong Kong.

Bayliss, R.F. (2002). 'Partnering on MTR Corporation Ltd's Tseung Kwan O Extension.' The Hong Kong Institution of Engineers Transactions, 9(1), 1–6.

Bayliss, R., Cheung S.O., Suen, H.C.H., and Wong, S.P. (2004). 'Effective partnering tools in construction: A case study on MTRC TKE contract 604 in Hong Kong.' *International Journal of Project Management, 22*(3), 253–263.

Bayramoglu, S. (2001). 'Partnering in construction: improvement through integration and collaboration.' *Leadership and Management in Engineering, ASCE,* 1(3), 39–43.

Beach, R., Webster, M., and Campbell, K.M. (2005). 'An evaluation of partnership development in the construction industry.' *International Journal of Project Management,* 23(8), 611–621.

Beamish, P.W. (1985). 'The characteristics of joint ventures in developed and developing countries.' *Columbia Journal of World Business,* 20(3), 13–19.

Beamish, P.W., and Delios, A. (1997). 'Incidence and propensity of alliance formation', In Beamish, P.W. and Killing, P.J. (eds), 'Co-operative strategies: Asian Pacific perspectives.' New Lexington Press, New York, 91–113.

Bellard, R.B. (1996). 'The partnering philosophy – a procurement strategy for satisfaction through a team work solution to project quality.' *Journal of Construction Procurement,* 2(1), 41–55.

Bennett, J. and Jayes, S. (1995). '*Trusting the team.*' Centre for Strategic Studies in Construction, The University of Reading.

Bennett, J., and Jayes, S. (1998). '*The seven pillars of partnering: A guide to second generation partnering.*' Thomas Telford Publishers, London.

Birnie, J. (1999). 'Private finance initiative (PFI) – UK construction industry response.' *Journal of Construction Procurement,* 5(1), 5–14.

Black, C., Akintoye, A., and Fitzgerald, E. (2000). 'An analysis of success factors and benefits of partnering in construction.' *International Journal of Project Management,* 18(6), 423–434.

Bloom, M.J. (1997). '*Partnering – A better way for doing business.*' The MITRE Corporation, Bedford, MA.

Bourn, J. (2001). '*Modernising Construction.*' National Audit Office, The Comptroller and Auditor General, London.

Bresnen, M., and Marshall, N. (2000a). 'Motivation, commitment and the use of incentives in partnerships and alliances.' *Construction Management and Economics,* 18(5), 587–598.

Bresnen, M., and Marshall, N. (2000b). 'Building partnerships: case studies of client-contractor collaboration in the UK construction industry.' *Construction Management and Economics,* 18(7), 819–832.

Bresnen, M., and Marshall, N. (2000c). 'Partnering in construction: a critical review of issues, problems and dilemmas.' *Construction Management and Economics,* 18(2), 229–237.

British Columbia. (1999). '*Public Private Partnership – A guide for local government*' Ministry of Municipal Affairs, Victoria, BC.

Brooke, K.L., and Litwin, G.H. (1997). 'Mobilizing the partnering process.' *Journal of Management in Engineering, ASCE,* 13(1), 42–50.

Brown, J. (1994). 'Partnering to save troubled projects.' *Journal of Management in Engineering, ASCE,* 10(3), 22–25.

Bubshait, A.A., and Almohawis, S.A. (1994). 'Evaluating the general conditions of a construction contract.' *International Journal of Project Management,* 12(3), 133–135.

Carrillo, P. (1996) 'Technology transfer on joint venture projects in developing countries.' *Construction Management and Economics,* 14(1), 45–54.

Chadwick, T., and Rajagopal, S. (1995). '*Strategic supply management.*' London: Butterworth-Heinemann, 92–117.

Chan, A.P.C. (1996). '*Determinants of project success in the construction industry of Hong Kong.*' Unpublished PhD thesis, University of South Australia.

Chan, A.P.C., and Chan, A.P.L. (2004). 'Key Performance Indicators (KPIs) for measuring construction success.' *Benchmarking: An International Journal,* 11(2), 203–221.

Chan, A.P.C., Chan, D.W.M., Fan, L.C.N., Lam, P.T.I., and Yeung, J.F.Y. (2004a). 'A Comparative Study of Project Partnering Practices in Hong Kong.' *Summary Report, Construction Industry Institute – Hong Kong,* Research Report No. 1, September.

Chan, A.P.C., Chan, D.W.M., Fan, L.C.N., Lam, P.T.I., and Yeung, J.F.Y. (2004b) 'A comparative study of project partnering practices in Hong Kong.' CD-Rom Full Report, Construction Industry Institute – Hong Kong, Research Report No. 2, April.

Chan, A.P.C., Chan, D.W.M., Fan, L.C.N., Lam, P.T.I., and Yeung, J.F.Y. (2004c). 'A Comparative Study of Project Partnering Practices in Hong Kong.' *Proceedings of the CII-HK Conference 2004 on Construction Partnering: Our Partnering Journey – Where Are We Now, and Where Are We Heading?*, 9 December 2004, Hong Kong, 65–75.

Chan, A.P.C., Chan, D.W.M., Chiang, Y.H., Tang, B.S., Chan, E.H.W., and Ho, K.S.K. (2004d). 'Exploring critical success factors for partnering in construction projects.' *Journal of Construction Engineering and Management, ASCE*, 130(2) 188–198.

Chan, D.W.M., Chan, A.P.C., and Yeung, J.F.Y. (2004e). 'Is structured approach a must for successful partnering venture? A case study of One Peking Road project in Hong Kong', *Proceedings of the CII-HK Conference 2004 on Construction Partnering: Our Partnering Journey – Where Are We Now, and Where Are We Heading?*, 9 December, Hong Kong, 97–103.

Chan, A.P.C., Chan, D.W.M., and Yeung, J.F.Y. (2004f). 'A Consultancy Report on the Interim Partnering Workshop for the Construction of the Residential Redevelopment at 26 Belcher's Street, Kennedy Town, Western District, Hong Kong (Ivy on Belcher's)' accepted by the client Hsin Chong Construction Co. Ltd., Hong Kong in March 2004.

Chan, A.P.C., Chan, D.W.M., Fan, L.C.N., Lam, P.T.I., and Yeung, J.F.Y. (2006a). 'Partnering for Construction Excellence – A Reality or Myth?', *Building and Environment*, 41(12), 1924–1933.

Chan, A.P.C., Chan, D.W.M., and Yeung, J.F.Y. (2006b). 'A Consultancy Report on the Initial Partnering Workshop for #136 The Hong Kong Community College Development at K.I.L. 11163 (Hung Hom Bay Campus)' accepted by the client Chevalier (Construction) Co. Ltd., Hong Kong in January 2006.

Chan, A.P.C., Chan, D.W.M., Fan, L.C.N., Lam, P.T.I., and Yeung, J.F.Y. (2008). 'Achieving partnering success through an incentive agreement: Lessons learned from an underground railway extension project in Hong Kong.' *Journal of Management in Engineering, ASCE*, 24(3), 119–195.

Chan, A.P.C., Chan, D.W.M., and Ho, K.S.K. (2002a). '*An analysis of project partnering in Hong Kong.*' Research Monograph, Department of Building and Real Estate, The Hong Kong Polytechnic University, Hong Kong, October.

Chan, A.P.C., Scott, D., and Lam, E.W.M. (2002b). 'Framework of success criteria for design/build projects.' *Journal of Management in Engineering, ASCE*, 18(3), 120–128.

Chan, A.P.C., Chan, D.W.M., and Ho, K.S.K. (2003a). 'An empirical study of the benefits of construction partnering in Hong Kong.' *Construction Management and Economics*, 21(5), 523–533.

Chan, A.P.C., Chan, D.W.M., and Ho, K.S.K. (2003b). 'Partnering in construction: Critical study of problems for implementation.' *Journal of Management in Engineering, ASCE*, 19(3), 126–135.

Chan, A.P.C., Cheng, E.W.L., and Li, H. (2001) '*Consultancy report on construction partnering in Hong Kong.*' The Hong Kong Housing Society.

Chan, E.H.W., and Suen, H.C.H. (2005). 'Disputes and dispute resolution systems in Sino-Foreign joint venture construction projects in China.' *Journal of Professional Issues in Engineering Education and Practice*, 131(2), 141–148.

Cheng, E.W.L., and Li, H. (2001). 'Development of a conceptual model of construction partnering.' *Engineering, Construction and Architectural Management*, 8(4), 292–303.

Cheng, E.W.L., and Li, H. (2002). 'Construction partnering process and associated critical success factors: Quantitative investigation.' *Journal of Management in Engineering, ASCE*, 18, 194–202.

Cheng, E.W.L. and Li, H. (2004a). 'A learning culture for strategic partnering in construction.' *Construction Innovation*, 4, 53–65.

Cheng, E.W.L., and Li, H. (2004b). 'Development of a practical model of partnering for construction projects.' *Journal of Construction Engineering and Management, ASCE*, 130(6), 790–798.

Cheng, E.W.L., Li, H., and Love, P.E.D. (2000). 'Establishment of critical success factors for construction partnering.' *Journal of Management in Engineering, ASCE*, 16(2), 84–92.

Cheng, E.W.L., Li, H., Lover, P., and Irani, Z. (2004). 'A learning culture for strategic partnering in construction.' *Construction Innovation*, 4(1), 53–65.

Cheung, S.O. (2001). 'Relationalism: Construction contracting under the People's Republic of China contract law.' *Cost Engineering*, 43(11), 38–43.

Cheung, S.O., Ng, T.S.T., Wong, S.P., and Suen, H.C.H. (2003a). 'Behavioural aspects in construction partnering.' *International Journal of Project Management*, 21(5), 333–343.

Cheung, S.O., Suen, H.C.H., and Cheung, K.K.W. (2003b). 'An automated partnering monitoring system – Partnering Temperature Index.' *Automation in Construction*, 12(3), 331–345.

Chow, G.C.H. (2002). *'Elements for Successful Implementation of Partnering in the Hong Kong Construction Industry.'* MSc Dissertation in Construction and Real Estate, Department of Building and Real Estate, The Hong Kong Polytechnic University.

Collin, J. (2002). *'Measuring the success of building projects – Improved project delivery initiatives.'* Queensland Department of Public Works, Australia.

Conley, M.A., and Gregory, R.A. (1999). 'Partnering on small construction projects.' *Journal of Construction Engineering and Management, ASCE*, 121(5), 320–324.

Construction Best Practice Programme. (1998). *'Lean Construction.'* Construction Best Practice Programme, Garston.

Construction Industry Board (CIB). (1997). *'Partnering in the team: A report by the Working Group 12 of the Construction Industry Board, UK.'* Thomas Telford, London.

Construction Industry Institute (CII). (1991). *'In search of partnering excellence.'* Publication no. 17–1, Report CII, Austin, TX.

Construction Industry Institute (Australia) (CII) (1996). *'Partnering: Models for success.'* Partnering Task Force, Construction Industry Institute, Australia.

Construction Industry Review Committee (CIRC) (2001). *'Construct for Excellence.'* Report by the Construction Industry Review Committee.' Hong Kong SAR Government, Hong Kong.

Contractor, F., and Lorange, P. (1988). *'Cooperative strategies in international business.'* Lexington Books, Boston, MA.

Cook, E.L. and Hancher, D.E. (1990). 'Partnering: contracting for the future.' *Journal of Management in Engineering, ASCE*, 6(4), 431–447.

Cowan, C., Gray, C., and Larson, E. (1992). 'Project partnering.' *Project Management Journal*, 22(4), 5–12.

Cox, R.F., Issa, R.R.A., and Ahrens, D. (2003). 'Management's perception of Key Performance Indicators for construction.' *Journal of Construction Engineering and Management*, 129(2), 142–151.

Crane, T.G., Felder, J.P. Thompson, P.J., Thompson, M.G., and Sanders, S.R. (1999). 'Partnering measures.' *Journal of Management in Engineering, ASCE*, 15(2), 37–42.

Crowley, L.G., and Karim, M.A. (1995). 'Conceptual model of partnering.' *Journal of Management in Engineering, ASCE*, 11(5), 33–39.

Currie and Brown. (2004). *'Do alliance projects offer value for money?'* Currie and Brown, Australia.

Demirbag, M., and Mirza, H. (2000). 'Factors affecting international joint venture success: An empirical analysis of foreign-local partner relationships and performance in joint ventures in Turkey.' *International Business Review*, 9(1), 1–35.

Department of Treasury and Finance. (2006). *'Project alliancing: Practitioners' guide.'* Department of Treasury and Finance, Melbourne, Australia.

Ding, D.Z. (1997). 'International partnering: a systematic framework for collaborating with foreign business partners.' *Journal of International Marketing*, 6(1), 91-107.

Dozzi, P., Hartman, F., Tidsbury, N., and Ashrafi, R. (1996). 'More stable owner-contractor relationships.' *Journal of Construction Engineering and Management, ASCE*, 122(1), 30–35.

Drexler, J. A., and Larson, E.W. (2000). 'Partnering: why project owner-contractor relationships change.' *Journal of Construction Engineering and Management, ASCE*, 126(4), 293–297.

Edelman, L., Carr, F., and Lancaster, C. (1991). *'Partnering, Army Engineer Institute for Water Resources.'* Fort Belvoir, VA.

Egan, J. (1998). *'The Egan Report – Rethinking Construction.'* Report of the Construction Industry Taskforce to the Deputy Prime Minister, HMSO, London.

Eisenberg, M.A. (2000). 'The emergence of dynamic contract law.' *California Law Review*, 88(6), 1743–1814.

Ellison, S. D., and Miller, D.W. (1995). 'Beyond ADR: working toward synergistic strategic partnering.' *Journal of Management in Engineering, ASCE*, 11(6), 44–54.

Environmental Protection Department. (2000). '*Review of the operation of Environmental Impact Assessment (EIA) Ordinance and the continuous improvement measures.*' Environmental Assessment and Noise Division, Environmental Protection Department, Hong Kong.

Fellows, R., and Liu, A. (2008). 'Research methods for construction.' 3rd Edition, Blackwell Science, Oxford: UK.

Florestano, P., and Gordon, S. (1980). 'Public vs. private: Small government contracting with the private sector.' *Public Administration Review*, 40(1), 29–34.

Gale, A., and Luo, J. (2004). 'Factors affecting construction joint ventures in China.' *International Journal of Project Management*, 22(1), 33–42.

Gardiner, P.D. and Simmon, J.E.L. (1998). 'Conflict in small- and medium-sized projects: case of partnering to the rescue.' *Journal of Management in Engineering, ASCE*, 14(1), 35–40.

Garvin, D.A. (1993). 'Building a learning organisation.' *Harvard Business Review*, 71(4), 78–91.

Gattorna, J.L., and Walters, D.W. (1996). '*Managing the supply chain.*' New York: Macmillan, 189–203.

Gentry, B., and Fernandez, L. (1997). 'Evolving public-private partnerships: General themes and urban water examples.' *Proceedings of the OECD Workshop on Globalisation and the Environment: Perspectives from OECD and Dynamic Non-Member Economies*, Paris, 13–14 November, 19–25.

Gerard, J. (1995). 'Construction.' *Journal of Construction Engineering and Management, ASCE*, 121(3), 319–328.

Geringer, J.M. (1988). '*Joint venture partner selection: Strategies for developed countries.*' Quorum Books, Westport, CT.

Geringer, J.M., and Hebert, L. (1989). 'Control and performance of international joint ventures.' *Journal of International Business Studies*, 20(2), 235–254.

Geringer, J. and Hebert, L. (1991). 'Measuring performance of international joint ventures.' *Journal of International Business Studies*, 22(2), 249–63.

Gerybadze, A. (1995). '*Strategic alliances and process redesign.*' Walter de Gruyter, New York.

Gibbons, R. (1996). '*Joint ventures in China: A guide for foreign investors.*' Macmillan Education Australia Pty Ltd, South Melbourne, Australia.

Goetz, C.J., and Scott, R.E. (1981). 'Principles of relational contracting.' *Virginia. Law Review*, 88(6), 1089–1150.

Gomes-Casseres, B. (1989). 'Joint ventures in the face of global competition.' *Sloan Management Review*, 30, Spring, 17–26.

Good, L. (1972). '*United States joint ventures and manufacturing firms in Monterrey, Mexico: Comparative styles of management.*' Unpublished PhD dissertation, Cornell University.

Gransberg, D.D., Dillon, W.D., Reynolds, L., and Boyd, J. (1999). 'Quantitative analysis of partnered project performance.' *Journal of Construction Engineering and Management, ASCE*, 125(3), 161–166.

Grant, T. (1996). 'Keys to successful public-private partnerships: General themes and urban water examples.' *Proceedings of the OECD Workshop on Globalisation and the Environment: Perspectives from OECD and Dynamic Non-Member Economies*, Paris, 13–14 November, 19–25.

Green, S.D. (1999). 'Partnering: The Propaganda of Corporatism?' *Journal of Construction Procurement*, 5(2), 177–186.

Gyles, R. (1992). '*The Royal commission into productivity in the building industry in New South Wales.*' New South Wales, Australia.

Hale, G. (1996). 'The leader's edge : mastering the five skills of breakthrough thinking.' Irwin Professional Publishers, Burr Ridge, IL.

Hampson, K.D., and Kwok, T. (1997). 'Strategic alliances in building construction: A tender evaluation tool for the public sector.' *Journal of Construction Procurement*, 3(1), 28–41.

Harback, H. F., Basham, D.L., and Buhts, R. E. (1994). 'Partnering paradigm.' *Journal of Management in Engineering, ASCE*, 10(1), 23–27.

Harrigan, K.R. (1985). 'Strategies for Joint Ventures.' Lexington Press: Lexington, MA.

Harrigan, K. (1986). '*Managing for Joint Ventures Success.*' Lexington Books, Lexington, MA.

Harrigan, K.R. (1987). 'Strategic alliance: Their new role in global competition.' *Columbia Journal of World Business*, 22(2), 67–69.

Harrigan, K.R. (2003). '*Joint ventures, alliances, and corporate strategy.*' Beard Books, Frederick, MD.

Hatush, Z., and Skitmore, M. (1997). 'Evaluating contractor prequalification data: Selection criteria and project success factors.' *Construction Management and Economics,* 15(2), 129–147.

Hauck, A.J., Walker, D.H.T., Hampson, K.D., and Peters, R.J. (2004). 'Project alliancing at National Museum of Australia – Collaborative Process.' *Journal of Construction Engineering and Management, ASCE,* 130(1), 143–152.

Hellard, R.B. (1996a). 'The partnering philosophy – a procurement strategy for satisfaction through a team work solution to project quality.' *Journal of Construction Procurement,* 2(1), 41–55.

Hellard, R.B. (1996b). '*Project partnering: principle and practice.*' Thomas Telford Publishers, London.

Herzfeld, E., and Wilson, A. (1996). '*Joint ventures (Third Edition).*' Jordan Publishing, Bristol.

Hewitt, I. (2001). '*Joint venture.*' Sweet and Maxwell, London.

Hinze, J. (1994). 'The contractor-subcontractor relationship: the subcontractor's view.' *Journal of Construction Engineering and Management, ASCE,* 120(2), 274–287.

Hipkin, I., and Naudé, P. (2006). 'Developing effective alliance partnerships: lessons from a case study.' *Long Range Planning,* 39(1), 51–69.

Hirsch, W., and Osborne, E. (2000). 'Privatisation of government services: pressure group resistance and service transparency.' *Journal of Labour Research,* 21(2), 315–326.

HM Treasury. (2000). 'Public private partnerships – the government's approach.' HMSO, London.

Hong Kong Efficiency Unit. (2002). '*Project 2002- Enhancing the Quality of Education in Glasgow City Schools by Public Private Partnership.*' HKSAR Government, Hong Kong.

Hong Kong Efficiency Unit. (2003). '*Serving the community by using the private sector – an introductory guide to public private partnerships (PPP).*' The Government of Hong Kong Special Administrative Region (HKSAR).

Hong Kong Efficiency Unit (2008). (http://www.eu.gov.hk/english/case/case.html)

Howarth, C., Gillin, M., Bailey, J. (1995). '*Strategic alliances: Resource-sharing strategies for smart companies.*' Pearson Professional (Australia) Pty Ltd, Australia.

Hu, M.Y., and Chen, H. (1996). 'An empirical analysis of factors explaining foreign joint venture performance in China.' *Journal of Business Research,* 35(1), 165–173.

Ireland, V. (1988). '*Improving work practices in the Australian building industry – a comparison with the Hong Kong and the USA. Master Builders.*' Report Commissioned by The Master Builders Federation of Australia, Sydney.

Jefferies, M., Gameson, R., and Rowlinson, S. (2002). 'Critical success factors of the BOOT procurement system: Reflection from the Stadium Australia case study.' *Engineering, Construction and Architectural Management,* 9(4), 352–361.

Jones, D. (2000). 'Project alliances.' *Proceedings of Conference on 'Whose risk? Managing risk in construction – who pays?',* Hong Kong, November.

Kanter, R.M. (1999). 'From spare change to real change.' *Harvard Business Review,* 77(2), 122–132.

Keating, M. (1998). 'Commentary: Public-private partnerships in the United States from a European perspective.' In Pierre, J. (ed): *Partnerships in urban governance: European and American Experience,* St Martin's Press, New York, 163–186.

Kelly, M. (1990). 'Strategic alliances.' Investing in Canada, Vol. 3, Spring, 1–3.

Kenny, A. (1975). '*Wittgenstein.*' Pelican Books, Harmondsworth.

Kilmann, R. (1995). 'A holistic program and critical success factors of corporate transformation.' *European Management Journal,* 13(2), 175–186.

Kogut, B. (1988). 'Joint ventures: theoretical and empirical perspectives.' *Strategic Management Journal,* 9, 319–332.

Kopp, J.C. (1997). '*Private capital for public works: Designing the next-generation franchise for Pubic-Private Partnerships in transportation infrastructure.*' Unpublished Master's thesis, Department of Civil Engineering, Northwestern University, Chicago, IL.

Kumaraswamy, M.M. (1997). 'Common categories and causes of construction claims.' *Construction Law Journal,* 13(1), 21–34.

Kumaraswamy, M.M., Love, P.E.D., Dulaimi, M., and Rahman, M. (2004). 'Integrating procurement and operational innovations for construction industry development.' *Engineering Construction and Architectural Management,* 11(5), 323–334.

Kumaraswamy, M.M. and Matthews, J.D. (2000). 'Improved subcontractor selection employing partnering principles.' *Journal of Management in Engineering, ASCE,* 16(3), 47–57.

Kumaraswamy, M.M., Rahman, M.M., Ling, F.Y.Y., and Phng, S.T. (2005). 'Reconstructing cultures for relational contracting.' *Journal of Construction Engineering and Management,* 131(10), 1065–1075.

Kwok, A., and Hampson, K. (1996). *'Building strategic alliances in construction.'* AIPM special publication, Queensland University of Technology.

Kwok, H.C., Then, D., and Skitmore, M. (2000). 'Risk management in Singapore construction joint ventures.' *Journal of Construction Research,* 1(2), 139–149.

Larson E. (1995). 'Project partnering: results of study of 280 construction projects.' *Journal of Management in Engineering, ASCE,* 11(3), 30–35.

Larson, E., and Drexler, J.A. (1997). 'Barriers to project partnering: report from the firing line.' *Project Management Journal,* 28(2), 46–52.

Latham, M. (1994). *'Constructing the team: Joint review of procurement and contractual arrangements in the United Kingdom construction industry.'* HMSO, London.

Lazar, F.D. (1997). 'Partnering – new benefits from peering inside the black box.' *Journal of Management in Engineering, ASCE,* 13(6), 75–83.

Lazar, F.D. (2000). 'Project partnering: improving the likelihood of win/win outcomes.' *Journal of Management in Engineering,* 16(2), 71–83.

Lecraw, D. (1983). 'Performance of transnational corporations in less developed countries.' *Journal of International Business Studies,* 14(1), 15–33.

Lendrum, T. (2000). *'The strategic partnering handbook – The practitioner's guide to partnerships and alliances'* (3rd edn). McGraw Hill, Sydney.

Lewis, J.D. (1995). *'The connected corporation.'* Free Press, New York, 1–13.

Li, B., and Akintoye, A. (2003). 'An overview of public-private partnership.' In Akintoye, A., Beck, M., and Hardcastle, C. (eds) *Public-private partnerships: Managing risks and opportunities,* Blackwell Science, Oxford, 1–24.

Li, B., Akintoye, A., Edwards, P.J., and Hardcastle, C. (2005). 'Critical success factors for PPP/PFI projects in the UK construction industry.' *Construction Management and Economics,* 23(5), 459–471.

Li, H., Cheng, E.W.L, and Love, P.E.D. (2000). 'Partnering research in construction.' *Engineering, Construction and Architectural Management,* 7(1), 76–92.

Li, H., Cheng, E.W.L., Love, P.E.D., and Irani, Z. (2001). 'Co-operative benchmarking: a tool for partnering excellence in construction.' *International Journal of Project Management,* 19 (3), 171–179.

Ling, F.Y.Y., Rahman, M.M., and Ng, T.L. (2006). 'Incorporating contractual incentives to facilitate relational contracting.' *Journal of Professional Issues in Engineering Education and Practice,* 132(1), 57–66.

Loraine, R.K. (1994). 'Project specific partnering.' *Engineering, Construction and Architectural Management,* 1(1), 5–16.

Love, S. (1997). 'Subcontractor partnering: I'll believe it when I see it.' *Journal of Management in Engineering, ASCE,* 13(5), 29–31.

Love, P.E.D., and Gunasekaran, A. (1999). 'Learning alliances: A customer-supplier focus for continuous improvement in manufacturing.' *Industrial and Commercial Training,* 31(3), 88–96.

Luo, K. (2001). 'Assessment management and performance of Sino-foreign construction joint ventures.' *Construction Management and Economics,* 19(2), 109–117.

Lyles, M.A., and Salk, J.E. (1996). 'Knowledge acquisition from foreign parents in international joint ventures: an empirical examination in the Hungarian context.' *Journal of International Business Studies, Special Issue,* 877–903.

Macaulay, S. (1963). 'Non-contractual relations in business: A preliminary study.' *American Sociological Review,* 28(1), 55–67.

MacBeth, D.K., and Ferguson, N. (1994). *'Partnership sourcing.'* Pitman, London, 96–140.

Macneil, I.R. (1974). 'The many futures of contracts.' *Southern California Law Review,* 47(2), 691–816.

Macneil, I.R. (1978). 'Contracts: Adjustment of long-term economic relations under classical, neoclassical and relational contract law.' *Northwestern University Law Review*, 72(5), part 2, 854–905.

Macneil, I.R. (1980). *'The new social contract: An inquiry into modern contractual relations.'* Yale University Press, New Haven, CT.

Madhok, A. (1995). 'Revisiting multinational firms' tolerance for joint ventures: a trust-based approach.' *Journal of International Business Studies*, 26, 117–137.

Main Roads Project Delivery System. (2005). *'Selection of Delivery Options (Volume One).'* Capability and Delivery Division Road System and Engineering, Department of Main Roads, Brisbane, Australia.

Manley, K., and Hampson, K. (2000). *'Relationship contracting on construction projects.'* QUT/CSIRO Construction Research Alliance, School of Construction Management and Property, Queensland University of Technology, Brisbane, Australia.

Martin, J. (2003). 'Performance measurement of time and cost predictability.' Working paper, FIG Working Week, Paris, France.

Mason, J.R. (2007). 'The views and experiences of specialist contractors on partnering in the UK.' *Construction Management and Economics*, 25(5), 519–527.

Mass Transit Railway Corporation Ltd. (MTRC). (2003a). *'The Tseung Kwan O Extension Success Story.'* Hong Kong.

Mass Transit Railway Corporation Ltd. (MTRC). (2003b). *'Tseung Kwan O Extension (TKE) influence of partnering on project cost.'* Working paper, Hong Kong.

Matthews, J. (1999). 'Applying partnering in the supply chain.' In Rowlinson, S. and McDermott, P. (eds). *Procurement Systems: A Guide to Best Practice in Construction*, E&FN Spon, London, 252–275.

Matthews, J., and Rowlinson, S. (1999). 'Partnering: incorporating safety management.' *Engineering, Construction, and Architectural Management*, 6(4), 247–257.

Matthews, J., Tyler, A., and Thorpe, A. (1996). 'Pre-construction project partnering: developing the process.' *Engineering, Construction and Architectural Management*, 3(1, 2), 117–131.

McCarthy, S.C., and Tiong, R.L.K. (1991). 'Financial and contractual aspects of build-operate-transfer projects.' *International Journal of Project Management*, 9(4), 222–227.

McGeorge, D. and Palmer, A. (2002). *'Construction management new directions'* (2nd edn). Blackwell Science, Oxford.

McInnis, A. (2000). *'Review on International Conference on "Whose risk? Managing risk in construction – who pays?"'* *Asian Architect & Contractor*, 29(11), 50–51.

McLennan, A. (2000). *'Relationship contracting: The main roads' perspective.'* Refereed paper presented to the Government Officials' Conference.

Mohr, J. and Spekman, R. (1994). 'Characteristics of partnering success: partnering attributes, communication behaviour, and conflict resolution techniques.' *Strategic Management Journal*, 15(1), 135–152.

Moore, C., Mosley, D., and Slagle, M. (1992). 'Partnering guidelines for win-win project management.' *Project Management Journal*, 22(1), 18–21.

Moore, D.R. (2000). 'Work-group communication patterns in design and build project teams: an investigative framework.' *Journal of Construction Procurement*, 6(1), 44–54.

Murphy, M.A. (1991). 'No more "What is communication?".' *Communication Research*, 18(6), 825–835.

Naoum, S.G. (1994). 'Critical analysis of time and cost of management and traditional contracts.' *Journal of Construction Engineering and Management*, 120(4), 687–705.

Naoum, S. (2003). 'An overview into the concept of partnering.' *International Journal of Project Management*, 21(1), 71–76.

National Health Service (NHS). (1999). *'Public private partnerships in National Health Service. The private financial service: Good practice.'* HMSO, London.

National Public Works Conference/National Building and Construction Council (NPWC/NBCC Report). (1990). *'Strategies for improvement in the Australian building and construction industry (No Dispute).'* Report by NPWC/NBCC Joint Working Party, May, Australia.

New South Wales Government. (2006). *'Working with Government - Guidelines for Privately Financed Projects.'* December.

Ng, L. (1997). 'Construction joint venture in Hong Kong.' In: Procurement – a key to innovation, *CIB Proceedings, Publication 203*, 535–544.

Ng, S.T., Rose, T.M., Mak, M., and Chen, S.E. (2002). 'Problematic issues associated with project partnering – the contractor perspective.' *International Journal of Project Management*, 20, 437–449.

Nicholson, G. (1996). 'Choosing the right partner for your joint venture, Fletcher constructions.' *Proceedings of the Joint Venture & Strategic Alliance Conference*. Sydney, Australia.

Nielsen, D. (1996). 'Partnering for performance.' *Journal of Management in Engineering, ASCE*, 12(3), 17–19.

Norris, W.E. (1990). 'Margin of profit: Teamwork.' *Journal of Management in Engineering, ASCE*, 6(1), 20–28.

Nyström, J. (2005). 'The definition of partnering as a Wittgenstein family-resemblance concept.' *Construction Management and Economics*, 23(5), 473–481.

Ofori, G. (1991). 'Programmes for improving the performance of contracting firms in developing countries: A review of approaches and appropriate options.' *Construction Management and Economics*, 9(1), 19–38.

Ozorhon, B., Arditi, D., Dikmen, I., and Birgonul, M.T. (2008). 'Implications of culture in the performance of international construction joint venture.' *Journal of Construction Engineering and Management*, 134(5), 261–370.

Parkhe, A. (1996). 'International joint ventures.' In *Handbook of International Management Research*, Punnett, J. and Shenkam, D. (eds), Blackwell, Oxford, 429–439.

Palaneeswaran, E., Kumaraswamy, M., Rahman, M., and Ng, T. (2003). 'Curing congenital construction industry disorders through relationally integrated supply chains.' *Building and Environment*, 38(4), 571–582.

Parkhe, A. (1991). 'Inter-firm diversity, organizational learning and longevity in global strategic alliances.' *Journal of International Business Studies*, 22, 579–601.

Parliament House Construction Authority (PHCA). (1990). 'Project Parliament – The management experience.' Parliament House Construction Authority, March, Australia.

Pena-Mora, F., and Harpoth, N. (2001). 'Effective partnering in innovative procured multicultural project.' *Journal of Management in Engineering, ASCE*, 17(1), 2–13.

Peters, R., Walker, D., and Hampson, K. (2001). '*Case study of the Acton Peninsula Development*.' Research and case study of the construction of the National Museum of Australia and the Australian Institute of Aboriginal and Torres Strait Islander Studies. School of Construction Management and Property, Queensland University of Technology.

Pfeffer, J., and Salancik, G. (1978). '*The external control of organizations: a resource dependence perspective*.' Harper and Row, New York.

Prokesch, S.E. (1997). 'Unleashing the power of learning: an interview with British Petroleum's John Browne.' *Harvard Business Review*, September-October, 147–168.

Qiao, L., Wang, S.Q., Tiong, R.L.K., and Chan, T.S. (2001). 'Framework for critical success factors of BOT projects in China.' *Journal of Project Finance*, 7(1), 53–61.

Queensland Department of Main Roads. (2005). '*Alliance contract (vol. 5)*.' Capability and Delivery Division, Department of Main Roads, Brisbane, Australia.

Rahman, M.M., and Kumaraswamy, M.M. (2002a). 'Risk management trends in the construction industry: Moving towards joint risk management.' *Engineering Construction and Architectural Management*, 9(2), 131–151.

Rahman, M.M., and Kumaraswamy, M.M. (2002b). 'Joint risk management through transactionally efficient relational contracting.' *Construction Management and Economics*, 20(1), 45–54.

Rahman, M.M., and Kumaraswamy, M.M. (2004). 'Potential for implementing relational contracting and joint risk management.' *Journal of Management in Engineering*, 20(4), 178–189.

Rahman, M.M., and Kumaraswamy, M.M. (2005). 'Relational selection for collaborative working arrangements.' *Journal of Construction Engineering and Management*, 131(10), 1087–1098.

Rahman, M.M., Kumaraswamy, M.M., and Ling, F.Y.Y. (2007). 'Building a relational contracting culture and integrated teams.' *Canadian Journal of Civil Engineering*, 34, 75–88.

Raneberg, D. (1994). 'Innovations in the public-private provision of infrastructure in the Australian state of New South Wales.' In: *Public Management – New Ways of Managing Infrastructure Provision*, OECD, Paris, 27–54.

Robson, M.J., and Dunk, M.A.J. (1999). 'Developing a pan-European co-marketing alliance: the case of BP-Mobil.' *International Marketing Review*, 16(3), 216–230.

Romancik, D.J. (1995). 'Partnership toward improvement.' *Project Management Journal*, 26(4), 14–20.

Ross, J. (2001). 'Introduction to project alliancing.' *Defence Partnering & Alliances Conferences*, Canberra, Australia, November.

Ross, J. (2003). '*Project alliancing – A strategy for avoiding and overcoming adversity.*' World Project Management Week, 24–28 March, Gold Coast, Australia.

Rowlinson, S., and Cheung, F.Y.K. (2004a). 'Relational contracting, culture and globalisation.' Proceedings of the CIB W107/TG23 Symposium on *Globalization and Construction*, November, Bangkok, Thailand.

Rowlinson, S., and Cheung, F.Y.K. (2004b). 'A review of the concepts and definitions of the various forms of relational contracting.' *Proceedings of the International Symposium of the CIB W92 on Procurement Systems 'Project Procurement for Infrastructure Construction'*, 7–10 January, Chennai, India.

Rowlinson, S., and Cheung, F.Y.K. (2004c). '*Relationship management in QDMR.*' RS&D Technical Forum in Bardon, Queensland, Australia.

Rowlinson, S., Cheung, F.Y.K., Simons, R., and Rafferty, A. (2006). 'Alliancing in Australia – no litigation contracts: A tautology?' *Journal of Professional Issues in Engineering Education and Practice*, 132(1), 77–81.

Ruff, C.M., Dzombak D.A., and Hendrickson, C.T. (1996). 'Owner-contractor relationships on contaminated site remediation projects.' *Journal of Construction Engineering and Management, ASCE*, 122(3), 348–353.

Sanders, S.R., and Moore, M.M. (1992). 'Perceptions on partnering in the public sector.' *Project Management Journal*, 22(4), 13–19.

Schultzel, H.J. (1996). '*Successful partnering: fundamentals for project owners and contractors.*' Wiley, New York.

Shaughnessy, H. (1995). 'International joint ventures: Managing successful collaborations.' *Long Range Planning*, 28(3), 1–9.

Skues, D. (1996). '*Partnering and its relevance to Hong Kong.*' Report by Crow Maunsell Management Consultants Ltd., Hong Kong.

Slater, T. S. (1998). 'Partnering: Agreeing to agree.' *Journal of Management In Engineering*, 14(6), 48–50.

Smith, A.J., and Walker, C.T. (1994). 'BOT: Critical factors for success.' A paper submitted at the '*Investment Strategies and Management of Construction*' *Conference*, September 20–24, Brijuni, Croatia, 247–254.

Smith, G. (1988). '*The report to the Full Bench of the Inquiry into the Building and Construction Industry.*' September.

Songer, A.D., and Molenaar, K.R. (1997). 'Project characteristics for successful public sector design/build.' *Journal of Construction Engineering and Management*, 123(1), 34–40.

Sridharan, G. (1995). '*Determinants of the JV success in Singapore construction industry.*' Unpublished PhD thesis, University College London.

Stonehouse, J.H., Hudson, A.R., and O'Keefe, M.J. (1996). 'Private-public partnerships: the Toronto Hospital experience.' *Canadian Business Review*, 23(2), 17–20.

Stephenson, R.J. (1996). '*Project partnering for the design and construction industry.*' Wiley, New York.

Tang, W., Duffield, C.F., and Young, D.M. (2006). 'Partnering mechanism in construction: an empirical study on the Chinese construction industry.' *Journal of Construction Engineering and Management, ASCE*, 132, 217–229.

Taylor, C.J. (1992). 'Ethyl Benzene project: the client's perspective.' *International Journal of Project Management*, 10(3), 175–178.

The International Project Finance Association. (2001). '*What are the benefits of PFI?*' The International Project Finance Association, London.

The KPI Working Group (2000). '*KPI report for the minister for construction.*' Department of the Environment, Transport and the Regions, London.

Thompson, P.J., and Sanders, S.R. (1998). 'Partnering continuum.' *Journal of Management in Engineering, ASCE*, 14(5), 73–78.

Thorpe, D. and Dugdale, G. (2004). 'Procurement and risk sharing: Use of alliance contracting for delivering local government engineering projects.' *Clients driving Innovation* International conference, Australia.

Tiong, R.L.K. (1996). 'CSFs in competitive tendering and negotiation model for BOT projects.' *Journal of Construction Engineering and Management, ASCE,* 122(3), 205–211.

Tiong, R.L.K., Yeo, K.T., and McCarthy, S.C. (1992). 'Critical success factors in winning BOT contracts.' *Journal of Construction Engineering and Management, ASCE,* 118(2), 217–228.

Tomlinson, J.W.C. (1970). '*The joint venture process in international business: India and Pakistan.*' MIT Press, Cambridge, MA.

Tsing Ma Bridge information (http://en.wikipedia.org/wiki/Tsing_Ma_Bridge) and (http://www.cse.polyu.edu.hk/~ctbridge/case/tsingma.htm).

Turner, R. (2002). 'Project success criteria.' *Project Magazine of the Association for Project Management,* 14(10), 32–33.

United States Trade Centre. (1998). http://ustradecenter.com/alliance.html#introduction.

Varadarajan, P.R., and Rajaratnam, D. (1986). 'Symbiotic marketing revisited.' *Journal of Marketing,* 50, January, 7–17.

Wakeman, T.H. (1997). 'Building sustainable public policy decisions through partnerships.' *Journal of Management in Engineering, ASCE,* 13(3), 40–47.

Walker, A. (1989). '*Project management in construction.*' BSP Professional Books, Oxford.

Walker, A., and Chau, K.W. (1999). 'The relationship between construction project management theory and transaction cost economics.' *Engineering Construction and Architectural Management,* 6(2), 166–176.

Walker, D.H.T, Hampson, K.D, Peters, R.J. (2000a). 'Project alliancing and project partnering – What's the difference? Partner selection on the Australian National Museum Project – A case study.' In: Serpell, A (ed). *Proceedings of CIBW92 Procurement System Symposium on Information and Communication in Construction Procurement,* Santiago, Chile, 641–655.

Walker, D.H.T, Hampson, K.D, and Peters, R. (2000b). '*Relationship-based procurement strategies for the 21st Century.*' RMIT University, Melbourne.

Walker D.H.T, Hampson, K, and Peters, R. (2002). 'Project alliancing vs project partnering: a case study of the Australian National Museum Project.' *Supply Chain Management: An International Journal,* 7(2), 83–91.

Walker, D.H.T., and Johannes, D.S. (2003). 'Construction industry joint venture behaviour in Hong Kong – designed for collaborative results?' *International Journal of Project Management,* 21(1), 39–49.

Walter, M. (1998). 'The essential accessory.' *Construction Manager,* 4(1), 16–17.

Weingardt, R.G. (1996). 'Partnering: building a stronger design team.' *Journal of Architectural Engineering, ASCE,* 2(2), 49–54.

Weston, D., and Gibson, G. (1993). 'Partnering project performance in US Army Corps of Engineers.' *Journal of Management in Construction,* 9(4), 331–344.

Whiteley, A., McCabe, M., Lawson, S. (1998). 'Trust and communication development needs an Australian waterfront study – An Australian waterfront study.' *Journal of Management Development,* 17, 432–446.

William, R.G., and Lilley, M.M. (1993). 'Partner selection for joint venture agreement.' *International Journal of Project Management,* 11(4), 233–237.

Wilson, R.A., Songer, A.D., and Diekmann, J. (1995). 'Partnering: more than a workshop, a Catalyst for change.' *Journal of Management in Engineering, ASCE,* 9(4), 410–425.

Wong, P.S.P., and Cheung, S.O. (2004). 'Trust in construction partnering: Views from parties of the partnering dance.' *International Journal of Project Management,* 22(6), 437–446.

Woodrich, A.M. (1993). 'Partnering: providing effective project control.' *Journal of Management in Engineering, ASCE,* 9(2), 136–141.

World Bank. (1986). '*The construction industry: Issues and strategies in developing countries.*' World Bank, Washington, DC.

Xu, T., Smith, N.J., and Bower, D.A. (2005). 'Forms of collaboration and project delivery in Chinese construction markets: probable emergence of strategic alliances and design/build.' *Journal of Management in Engineering, ASCE,* 21(2), 100–109.

Yan, A., and Luo, Y. (2001). 'International joint ventures: theory and practice.' M.E. Sharpe, Armonk, NY.

Yeong, C.M. (1994). 'Time and cost performance of building contracts in Australia and Malaysia.' MSc thesis, University of South Australia, Australia.

Yeung, J.F.Y., Chan, A.P.C., and Chan, D.W.M. (2007a). 'The definition of alliancing in construction as a Wittgenstein family-resemblance concept.' *International Journal of Project Management*, 25(3), 219–231.

Yeung, J.F.Y., Chan, A.P.C., Chan, D.W.M., and Leong-kwan Li (2007b). 'Development of a partnering performance index (PPI) for construction projects in Hong Kong: a Delphi study.' *Construction Management and Economics*, 25(12), 1219–1237.

Yeung, J.F.Y., Chan, A.P.C., and Chan, D.W.M. (2009). 'Development of a performance index (PI) for relationship-based construction projects in Australia: a Delphi study.' *Journal of Management in Engineering, ASCE*, 25(2), 1–10.

Zhang, W.R., Wang, S.Q., Tiong, R.L.K., Ting, S.K., and Ashley, D. (1998). 'Risk management of Shanghai's privately financed Yan'an Donglu tunnels.' *Engineering, Construction and Architectural Management*, 5(4), 399–409.

Zhao, F. (2002). 'Measuring Inter-organizational Partnership: The Challenge of Cultural Discrepancy.' *Proceedings of the 3rd International Conference on Theory & Practice in Performance Measurement*, Boston, MA.

Index